NATURAL LANGUAGE PROCESSING IN ARTIFICIAL INTELLIGENCE

Edited by
Brojo Kishore Mishra, PhD
Raghvendra Kumar, PhD

APPLE
ACADEMIC
PRESS

Apple Academic Press Inc.
4164 Lakeshore Road
Burlington ON L7L 1A4, Canada

Apple Academic Press Inc.
1265 Goldenrod Circle NE
Palm Bay, Florida 32905, USA

First issued in paperback 2021

© 2021 by Apple Academic Press, Inc.

Exclusive co-publishing with CRC Press, a Taylor & Francis Group

No claim to original U.S. Government works
ISBN-13: 978-1-77188-864-6 (hbk)
ISBN-13: 978-1-77188-865-3 (pbk)
ISBN-13: 978-0-36780-849-5 (eBook)

Library and Archives Canada Cataloguing in Publication

Title: Natural language processing in artificial intelligence / edited by Brojo Kishore Mishra, PhD, Raghvendra Kumar, PhD.

Names: Mishra, Brojo Kishore, 1979- editor. | Kumar, Raghvendra, 1987- editor.

Description: Includes bibliographical references and index.

Identifiers: Canadiana (print) 20200224727 | Canadiana (ebook) 20200224816 | ISBN 9781771888646 (hardcover) | ISBN 9780367808495 (ebook)

Subjects: LCSH: Natural language processing (Computer science) | LCSH: Artificial intelligence.

Classification: LCC QA76.9.N38 N38 2020 | DDC 006.3/5—dc23

Library of Congress Cataloging-in-Publication Data

Names: Mishra, Brojo Kishore, 1979- editor. | Kumar, Raghvendra, 1987- editor.

Title: Natural language processing in Artificial Intelligence / edited by Brojo Kishore Mishra, PhD., Raghvendra Kumar, PhD.

Description: Burlington, ON, Canada ; Palm Bay, Florida : Apple Academic Press, 2020. | Includes bibliographical references and index. | Summary: "This volume, Natural Language Processing in Artificial Intelligence, focuses on natural language processing (NLP), artificial intelligence (AI), and allied areas. The book delves into natural language processing, which enables communication between people and computers and automatic translation to facilitate easy interaction with others around the world. It discusses theoretical work and advanced applications, approaches, and techniques for computational models of information and how it presented by language (artificial, human, or natural in other ways). It looks at intelligent natural language processing and related models of thought, mental states, reasoning, and other cognitive processes. It explores the difficult problems and challenges related to partiality, under specification, and context-dependency, which are signature features of information in nature and natural languages. Key features: Addresses the functional frameworks and workflow that are trending in NLP and AI Explores basic and high level concepts, thus serving as a resource for those in the industry while also helping beginners to understand both basic and advanced aspects Looks at the latest technologies and the major challenges, issues, and advances in NLP and AI Explores an intelligent field monitoring and automated system through AI with NLP and its implications for the real world Discusses data acquisition and presents a real-time case study with illustrations related to data-intensive technologies in AI and NLP Topics include the process of business intelligence and how this platform is used, the concepts of Information retrieval systems, the neural machine translation (NMT) process, the choice of words and text in natural language processing, embedded traffic control and management systems, a technique for generating ontology by adopting the fruit fly optimization algorithm, POS labeling using the Viterbi algorithm, how natural language processing techniques can be used to prevent phishing attacks, and more. This volume will be a useful and informative resource for faculty, advanced-level students, and professionals in the field of artificial intelligence, natural language processing, and other areas"-- Provided by publisher.

Identifiers: LCCN 2020016543 (print) | LCCN 2020016544 (ebook) | ISBN 9781771888646 (hardcover) | ISBN 9780367808495 (ebook)

Subjects: LCSH: Natural language processing (Computer science)

Classification: LCC QA76.9.N38 N3846 2020 (print) | LCC QA76.9.N38 (ebook) | DDC 006.3/5--dc23

LC record available at https://lccn.loc.gov/2020016543

LC ebook record available at https://lccn.loc.gov/2020016544

Apple Academic Press also publishes its books in a variety of electronic formats. Some content that appears in print may not be available in electronic format. For information about Apple Academic Press products, visit our website at **www.appleacademicpress.com** and the CRC Press website at **www.crcpress.com**

About the Editors

Brojo Kishore Mishra, PhD
Professor, Department of Computer Science and Engineering,
GIET University, Gunupur, Odisha, India

Brojo Kishore Mishra, PhD, is a Professor in the Computer Science and Engineering Department at the Gandhi Institute of Engineering and Technology University (GIET), Gunupur, Odisha, India. He has published more than 30 research papers in national and international conference proceedings, over 25 research papers in peer-reviewed journals, and over 22 book chapters, and has authored two books and edited three books to date. His research interests include data mining and big data analysis, machine learning, soft computing, and evolutionary computation. He received his PhD degree in Computer Science from the Berhampur University, Brahmapur, Odisha, India.

Raghvendra Kumar, PhD
Associate Professor, Computer Science & Engineering Department,
GIET University, Gunupur, Odisha, India

Raghvendra Kumar, PhD, is an Associate Professor in the Computer Science and Engineering Department at the Gandhi Institute of Engineering and Technology University (GIET), Gunupur, Odisha, India. He also serves as Director of the IT and Data Science Department at the Vietnam Center of Research in Economics, Management, Environment, Hanoi, Viet Nam. Dr. Kumar serves as Editor of the book series Internet of Everything: Security and Privacy Paradigm (CRC Press/Taylor & Francis Group) and the book series Biomedical Engineering: Techniques and Applications (Apple Academic Press). He has published a number of research papers in international journals and conferences. He has served in many roles for international and national conferences, including organizing chair for several conferences, volume editor, volume editor, keynote speaker, session chair or co-chair, publicity chair, publication chair, advisory board member, and technical program committee member. He has also served as a guest editor for many special issues of reputed journals. He authored and edited 17 computer science books in field of internet of things, data mining, biomedical engineering, big data, robotics, graph theory, and Turing machines. He is the Managing Editor of the *International Journal of Machine Learning and Networked Collaborative Engineering*. He received a best paper award at the IEEE Conference 2013 and Young Achiever Award–2016 by IEAE Association for his research work in the field of distributed database. His research areas are computer science-cloud computing, big data and database, security and privacy, multimedia system, machine learning, computational Intelligence, and image processing. Dr. Kumar received his BTech in Computer Science and Engineering from SRM University Chennai (Tamil Nadu), India, his MTech in Computer Science and Engineering from KIIT University, Bhubaneswar, (Odisha) India, and his PhD in Computer Science and Engineering from Jodhpur National University, Jodhpur (Rajasthan), India.

Contents

Contributors

Amiya Bhusan Bagjadab
Sambalpur University of Information Technology, Burla, India, E-mail: amiya7bhusan7@gmail.com

B. Balamurugan
Professor, School of Computing Science and Engineering, Galgotias University, Greater Noida, Uttar Pradesh, India, E-mail: kadavulai@gmail.com

Balamurugan Balusamy
School of Computer Science and Engineering, Galgotias University, Noida, Uttar Pradesh, India

Jyotir Moy Chatterjee
School of Computing Science and Engineering, Department of IT, LBEF (APUTI), Kathmandu, Nepal, E-mail: jyotirm4@gmail.com

Bishwa Ranjan Das
North Orissa University, Baripada, India, E-mail: biswadas.bulu@gmail.com

E. Ganga Devi
Loyola College, Chennai, Tamil Nadu, India

G. S. Pradeep Ghantasala
Professor, Department of Computer Science and Engineering, Malla Reddy Institute of Technology and Science, Hyderabad, Telengana, India, E-mail: ggspradeep@gmail.com

Soumitra Ghosh
Department of Computer Science and Engineering, Indian Institute of Technology Patna, India

Mohd Shahid Husain
Assistant Professor, Information Technology Department, College of Applied Sciences, Ibri, Ministry of Higher Education, Oman, Tel.: 968-94714261, E-mails: siddiquisahil@gmail.com, mshahid.ibr@cas.edu.om

Jeevanandam Jotheeswaran
School of Computing Science and Engineering, Galgotias University, Greater Noida, Uttar Pradesh, India, E-mail: jeevanandamj@gmail.com

S. Karthikeyan
School of Computing Science and Engineering, Galgotias University, Greater Noida, Uttar Pradesh, India, E-mail: link2karthikcse@gmail.com

Firoz Khan
Professor, IT Faculty, Dubai Mens College, Higher Colleges of Technology, UAE, E-mail: fk7@hotmail.com

Mishra Brojo Kishore
Department of Computer Science and Engineering, GIET University, Gunupur, Odisha, India, E-mail: brojokishoremishra@gmail.com

Chandan Koner
Dr. B. C. Roy Engineering College, Durgapur, West Bengal, India

Mishra Sambit Kumar
Department of Computer Science and Engineering, Gandhi Institute for Education and Technology, Bhubaneswar, Odisha, India, E-mail: sambitmishra@gietbbsr.com

Raghvendra Kumar
Associate Professor, Computer Science & Engineering Department, GIET University, India
E-mail: raghvendraagrawal7@gmail.com

Chandrakanta Mahanty
Department of CSE & IT, GIET University, Gunupur, Odisha, India

K. Mandal
Assistant Professor, School of Computing Science and Engineering, Galgotias University, Uttar Pradesh, India, E-mail: kuppanmandal@gmail.com

Brojo Kishore Mishra
Department of Computer Science and Engineering, & IT, GIET University, Gunupur, Odisha, India, E-mail: brojokishoremishra@gmail.com

Anirban Mitra
Amity University, Kolkata, West Bengal, India

Devpriya Panda
Department of CSE & IT, GIET University, Gunupur, Odisha, India

Subrata Paul
Rearch Scholar, MAKAUT, and Annex College, Kolkata, West Bengal, India, E-mail: subratapaulcse@gmail.com

Sushree Bibhuprada B. Priyadarshini
Institute of Technical Education and Research, Siksha 'O' Anusandhan (Deemed to be University), Bhubaneswar, India, E-mail: bimalabibhuprada@gmail.com

R. Sathiyaraj
Assistant Professor, School of Computing Science and Engineering, Galgotias University, Uttar Pradesh, India, E-mail: sathiya.peace@gmail.com

M. Lawanya Shri
School of Information Technology and Engineering, VIT, Vellore, India

Prasad Suman Sourav
Department of MCA, Ajay Binay Institute of Technology, Cuttack, Odisha, India, E-mail: prasadsuman800@rediffmail.com

Abbreviations

ADJ	adjective
ADV	adverb
AI	artificial intelligence
ARM	association rule mining
BLEU	beam size of resulting translations
BMS	building management systems
BN	Bayesian net
CA	content analysis
CBMT	corpus-based machine translation
CDMT	corpus-driven machine translation
CI	computational intelligence
CLRS	cross language retrieval systems
CONJ	conjunction
CRM	customer relationship management
DBMT	dictionary based machine translation
DDMT	data-driven machine translation
DeepNMT	deep neural machine translation
DM	data mining
DT	decision tree
EC	expectation confirmation
EN -> FR	English to French
FN	false negatives
FOA	fruit fly algorithm
FP	false positives
FURIA	fuzzy unordered rule induction
GA	genetic algorithm
GAM	generalized additive model
GBM	gradient boosting
GLM	generalized linear model
GNMT	Google neural machine translation
GRNN	generalized regression neural network
HAC	hierarchical agglomerative clustering
HMM	hidden Markov model

HMT	hybrid machine translation
IBk	instance-based learning
IE	information extraction
IR	information retrieval
KDD	knowledge discovery from data
KNN	K-nearest-neighbors
KPIs	key-performance indicators
LA	learning algorithm
LR	logistic regression
LSI	latent semantic indexing
MC	model combination
MKP	multidimensional knapsack problem
N	noun
NB	Naive Bayes
NER	named entity recognition
NLG	natural language generation
NLP	natural language processing
NLTK	natural language toolkit
NLU	natural language understanding
NMT	neural machine translation
NN	neural networks
NNOM	noun-nominative
NSOC	noun-sociative
OCR	optical character recognition
OpenNMT	open source neural machine translation
OWL	web ontology language
PAMIR	parallel multimedia information retrieval
PCA	principal component analysis
PREP	preposition
RDB	relational databases
RDF	resource description framework
RF	random forest
SBI	social business intelligence
SG	semantic graphs
SMART	specific, measurable, attainable, realistic, and time-sensitive
SMO	sequential minimal optimization
SMT	statistical machine translation
SVM	support vector machine

SWoT	semantic web of things
TF	term frequency
TM	text mining
TM	translation memory
TN	true negatives
TP	true positives
TRP	target rating point
V	verb
VFIN	verb finite
VIF	variability inflation factor
VNFIN	verb nonfinite
VSM	vector space demonstrate
WSD	word-sense disambiguation
WSN	wireless sensor networks
WWW	World Wide Web

Preface

Natural language processing (NLP) enables communication between people and computers and automatic translation to enable people to interact easily with others around the world. The extraordinary development of the internet and the explosion of textual data on the web have boosted the development of the natural language processing field and have especially led to the revival of corpus based NLP and linguistics. Computational and technological developments that incorporate natural language are proliferating. Adequate coverage encounters difficult problems related to partiality, under specification, and context-dependency, which are signature features of information in nature and natural languages. Furthermore, agents (humans or computational systems) are information conveyors, interpreters, or participate as components of informational content. Generally, language processing depends on agents' knowledge, reasoning, perspectives, and interactions.

The aim of this edited book is to foster interactions among researchers and practitioners in NLP, AI, and allied areas. The edited book covers theoretical work, advanced applications, approaches, and techniques for computational models of information and its presentation by language (artificial, human, or natural in other ways). The goal is to promote intelligent natural language processing (NLP) and related models of thought, mental states, reasoning, and other cognitive processes.

The book is organized into thirteen chapters. Chapter 1 presents a review of the entire process of business intelligence, and then brings out insights on how this platform is used in order to undertake decisions by means of social networks.

Chapter 2 deals with the basic concepts of information retrieval (IR) systems, their needs, models of information retrieval systems, and other related concepts, like stemming, indexing, etc. The chapter will help scholars and young professionals to get critical information required for developing IR systems.

Chapter 3 explains all forms of neural machine translation (NMT), with complete translation of a process that involves a neural network, which produces a number of accessible inputs to find the best possible output according to utilization. This kind of translation applies multiple strategies on different stages for translation. Stage one implements the translated based on words with a complete sentence, and Stage 2 implements the translated on a model over the word within the sentence context. The solidity of neural machine translation allows the learning ability over point-to-point bases on the background knowledge input to the predicable target output. The chapter also includes a brief discussion on various kinds of neural machine translation conversion principles, such as like Google neural machine translation (GNMT), open source neural machine translation (OpenNMT), deep neural machine translation (DeepNMT), and so on.

Chapter 4 discusses how natural language processing may be linked to accomplish human-like language processing. Choosing the word during performance may be very deliberate and may not be replaced with actual understanding. The complete perception of natural language processing may be associated with as making a phrase and outlined the input text and conversion of the text into similar language in different form, queries about the text, and inference from the text.

A revolution in the transport environment needs–redesigning of the infrastructure so that the production of embedded vehicles can be chained to an embedded traffic management system. This instinctual design of the traffic control and management system can lead to the improvement of the traffic congestion problem. The traffic density can be calculated using a Raspberry Pi microcomputer and a couple of ultrasonic sensors, and the lanes can be operated accordingly. A website can be designed where traffic data can be uploaded and any user can retrieve it. This property will be useful to users for getting real-time information and detection of any road intersection and discover the fastest traffic route.

Chapter 5 proposes an effective technique for generating ontology by adopting the fruitfly optimization algorithm (FOA). The proposed approach proves that the construction of ontology-based information increases the logistics of the system effectively with low operating cost.

Chapter 6 focuses on supervised approaches (rule based and stochastic-based). Under the stochastic approach, an elaborate discussion on POS labeling using the Viterbi algorithm has been done here. We have also discussed various popularly used POS taggers along with their implementation with proper examples to give a vivid idea on how they work.

Chapter 7 discusses the available methodologies to perform and the most familiar algorithms. The applications are not limited to the classification of research papers, healthcare, customer relationship management (CRM), education; it applies in all places where text has been stored as data. As it is already told that information is wealth; text mining (TM) supports using the information to the next level of decision making for business, result analysis for the education sector, and so on. The detailed text mining algorithms are discussed, such as decision trees, neural networks, discovering algorithms, differential evolution, and so on. Applications such as healthcare, banks, social media, customer relationship and all are connected to the text mining. These applications limited where as the usage of text mining are not limited to the above means. In simple terms, wherever the text has been stored as data in the database, it can be used extensively in taking decisions, predicting results, diagnosing patients, increasing sales, and so on.

Chapter 8 discusses the two basic aspects of natural language understanding (NLU) and natural language generation (NLG) that deals with natural language understanding and generation respectively. Further, the current chapter throws light on various aspects of text processing, like morphological analysis, syntax analysis, semantic analysis, lexical analysis, etc.

Chapters 9 discusses the types of phishing attacks. It also focuses on an anti-phishing URL tool, which is used to prevent phishing attacks. The main objective of this chapter is to explain initially the characteristics of phishing attacks. There are some uniqueness and patterns associated with the websites that are used for phishing. Their properties can be used to detect phishing. Then these attacks are detected by a hybrid machine learning model. The system has been implemented by examining the URLs used in phishing attacks with some extracted features before opening them. Some natural language processing techniques are used in the proposed machine learning system. These techniques are used for analyzing the text semantically to detect malicious intentions that indicate phishing attacks. In order to identify the websites for their legitimacy, some machine learning algorithms (LAs) are also discussed in this chapter. It also focuses on Naive Bayes (NB) classifier and K-Means clustering to calculate the possibility of the website as valid phish or invalid phish.

Chapter 10 discusses that natural language processing is subfield of artificial intelligence (AI) and a research area in the field of computer science recently. It is processed by the computer system and understands the concept that is given as text input and generates some meaningful result. There are many subfields of natural language processing, like machine translation, information retrieval, information extraction (IE), and question answering. There are different types of natural language are available in India, even if in worldwide label like Hindi, Odia, Bengali, Marathi, French, Spanish, and German etc.

We are sincerely thankful to Almighty for supporting and standing at all times with us, whether it's good or tough times and given ways to conceded us.

Starting from the call for chapters till the finalization of chapters, all the editors have given their contributions amicably, which it a positive sign of significant team works. The editors are sincerely thankful to all the members of Apple Academic Press, especially Sandra Jones Sickels for the providing constructive input and allowing opportunity to edit this important book. We are equally thankful to a reviewers who hail from different places in and around the globe and who shared their support and who stand firm towards quality chapters. The rate of acceptance we have kept as low as 16% to ensure the quality of work submitted by author.

The aim of this book is to support the computational studies at the research and post-graduation level with open problem solving techniques. We are confident that it will bridge the gap by supporting novel solutions to support them in their problem solving. At the end, the editors have the taken utmost care while finalizing the chapters for the book, but we are open to receive your constructive feedback, which will enable us to carryout necessary points in our forthcoming books.

—Brojo Kishore Mishra, PhD
Raghvendra Kumar, PhD

CHAPTER 1

A Survey on Social Business Intelligence: A Case Study of Application of Dynamic Social Networks for Decision Making

SUBRATA PAUL,[1] CHANDAN KONER,[2] and ANIRBAN MITRA[3]

[1]Research Scholar, MAKAUT, and Annex College, Kolkata, West Bengal, India, E-mail: subratapaulcse@gmail.com

[2]Dr. B C Roy Engineering College, Durgapur, West Bengal, India

[3]Amity University, Kolkata, West Bengal, India

ABSTRACT

Over the years, the popularity of social network platforms has increased alarmingly as they are castoff by people for expression of thoughts. Platforms are also used by firms for acknowledgment of numerous opportunities which would lead to the fulfillment of their objectives. The demand for information by firms brings out its fundamental inclinations and dependences that would largely affect performance of the firms. Business intelligence systems are used for obtaining these perceptions. The systems aimed in the derivation of actionable information from social media platforms for the provision of managerial decision-making are demonstrated as social business intelligence (SBI) systems. Numerous queries being raised ranging from ways firms process the external data, the management information derived from new data sources and successful implication of the SBI. In this chapter, we had presented a review of the entire process of business intelligence and then brought out insights on how this platform is used in order to undertake any decision by means of the social network. This work is purely a review paper.

1.1 INTRODUCTION

Social network enlightens the collection of associates amongst personages, where every personage is elaborated as a social entity. The assemblage of knots between communities with supremacy of those knots is established through social networks. Commencing the mathematical perspective, this social structure contains nodes of personages or organizations which are further narrated with distinct or compound-specific varieties of alliance regulations [1].

In addition, social networks can also have a dynamic behavior. Ties are assembled, which might flourish producing close association and silent liquefaction or unexpectedly revolve sour and perish further. Dynamic social networks might be fruitful in modeling and analysis of human associations in numerous possible circumstances including informal social relationship among personages surrounded by family or cluster of pals; the ordered alliance of employees in outsized endeavors; extensive correlations during social networking services; or conversion of activities of minute, interrelated terrorist cells [2].

Business intelligence can be described as a procedure of translation of data into information which would facilitate in managerial decision-making. Elbashir et al. [3] described that "business intelligence a system provides ability in analysis of the business information for the support and improvement of managerial decision making within a wide aspect of business activities." Van Beek [4] has characterized business intelligence as "an uninterrupted procedure which will help organization to gather and register data, analyze it with subsequent application of resultant information and knowledge towards decision-making procedure for the improvement of organizational presentation." Rouibah and Ould-Ali [5] further illustrated business intelligence as "a tactical approach for methodically boarding, pursuing, communication, and transformation of applicable pathetic indications hooked on actionable facts which formed the basis of premeditated decision-making." Even though these explanations differ marginally, the related characteristics lies on the fact that business intelligence is supposed as a procedure of translation of data into interpretable information that will finally lead to managerial decision-making.

The performance of an organization should be measurement criteria for calculation of the effects of social media activities. In order to measure the performance of a firm against its strategy, there lies a requirement of key performance indicator which should be defined by the process of business intelligence. This value-based management method is commonly functional inside firms which will suggest when they will follow to accomplish social media happenings that would count the shortcomings of these happenings in comparison with the performance of organization. Prevailing social media monitoring tools primarily disclose organizational performance on social media wherein its constituent on a firm is granted as a discrete business unit accomplishing its specific approach. Although the persistence of business intelligence lies on the exposure of the essential parameters which determine the performance of an organization, it is merely dependent on the presentation of social media. For assisting the authority of social media content towards the performance of organization, there lies a necessity of links among the organization's primary-performance indicators and clear social media constraints.

Frequently an argument is seen beneath potentials of social media towards business intelligence resolutions that influences supplementary amongst modules which are presently obtainable with social media methodical apparatus. Main assistances shall sum up every time a link is made between KPIs of organizations and their constraints which might be distinguished through social media apparatus. Within social business, intelligence there lays the social media happenings being associated with a firm which transforms in organizational presentation.

1.2   BUSINESS INTELLIGENCE PERSPECTIVES

Van Beek [4] pictures business intelligence ("BI") as a cycle, comprising of three key procedures:

1. **Register:** BI cycle commences sensibly by listening-*registration*—to environment. Surrounded by environments, dissimilarity is prepared amongst circumstantial besides transactional environments which comprises of features which has effects on the organization and actors having a straight association with company, comparable to customers, suppliers, employees, and competitors respectively.

2. **Process:** Accordingly, when data (in whatever format) is registered, there lies a requirement for its *processing* enabling the assembled statistics discloses tendencies with the delivery of valued facts.

Van Beek [4] sites 'small BI cycle' inside the procedure; statistics are collected, there is an analysis, assembling, and circulation till the correct administrative subdivisions.

3. **React:** Subsequently, after dispensation of statistics, the corporation is able in *reacting*. Van Beek [4] contends that a corporation might respond on three stages; functioning, strategic or intentional. Accordingly, the atmosphere assesses the corporations' dissimilarities in communications, ensuing in novel indications for the company's BI cycle.

1.2.1 TYPOLOGY OF PERFORMANCE INDICATORS

Performance indicators consist of vital components in business intelligence, because of facts which they imitate the presentation of the happenings that subsidize to the firm's strategy.

1. **Leading and Lagging Indicators:** There are dualistic important categories of indicators; leading indicators and lagging indicators. *Leading* indicators principals to outcomes which in addition are marked as '(value) drivers.' *Lagging* indicators are the significances which amounts on productivity of historical happenings, besides recognized as 'consequences' [6]. Leading indicators deal with the management, while lagging indicators provide a measurement on the intensity of management. With leading indicators, there lies a possibility of direct response during the arousal of low outcomes [6].

    Illustratively, Table 1.1 illustrates a few scenarios of leading and lagging indicators.

TABLE 1.1   Illustrations of Leading and Lagging Indicators [6]

Leading Indicators	Lagging Indicators
Pioneering sales currently	Proceeds
Considered rearticulate currently	Price
Customer possessions currently uncovered	Capability
Documented software pollutions	Customer consummation
	Worker conservation
	Restrictions
	Trustworthiness

2. **Quantitative and Qualitative Indicators:** Another distinction between metrics is the difference between quantitative or qualitative based indicators. The quantitative indicator uses counting, adding, and averaging, etc. for counting processes. Examples of quantitative measures are inventories, number of orders, number of clients, the delivery time of goods, sales, other financial figures, etc. In comparison to qualitative indicators, quantitative indicators are relatively easy to measure.

3. **Key Performance Indicators:** To distinguish amongst performance indicators that are more prime than others, some indicators are called 'key-performance indicators' (KPIs). According to Tsai and Cheng [7], KPIs are "the groundwork of the performance system which turns the strategic goals of a company into long-term objectives." Along with the word 'key,' a performance indicator provides an indication of being more attentive. Table 1.2 lists the constituents which a key-performance indicator should fulfill, namely specific, measurable, attainable, realistic, and time-sensitive (SMART) [8].

TABLE 1.2 Requirements of a Key-Performance Indicator [8]

Requirement	Description
Precise	KPIs must be comprehensive besides as precise as conceivable.
Quantifiable	A KPI should be quantifiable in contradiction of typical presentation and a customary of anticipation.
Achievable	The aim of a KPI should be concentrating. They should be sensible and achievable.
Truthful	A objective should be accurate in agreement with the exact employed situation
Time delicate	Objectives should be accomplished inside a time surround for observing the development.

1.2.2 BI PROCESSING: FROM DATA TO INFORMATION

There is a requirement of processing raw data by registering signal results prior to representing information. In the second phase of BI, there is a processing of registered signals. A social media constituent should have consistency with the existing system(s) and process(es). Van Beek [4] distinct 15 activities which make up the processing of gathered data into information. Figure 1.1 shows the activities in the BI cycle.

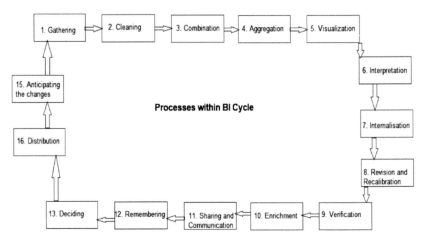

FIGURE 1.1   Activities of BI cycle [4].

The fifteen steps are discussed in the following section:

1. **Collecting:** The stored data from different system in form of signals are collected in a separate system. In contrast with social business intelligence (SBI), there is a requirement of different social media platforms-like Twitter and Facebook for data collection. Every social media platform has a self driven methodology for data storage.
2. **Filtering:** The signals from which meaningful information is deduced pass the filtering process. The outdated or inferior quality data are isolated. This process is important when we use social media data for BI purposes. The data may constitute of spam which pollutes the data.
3. **Combine:** Combination and Integration of the collected and filtered data from separate systems into a single source is carried out.
4. **Aggregating:** Aggregation of detailed data up to a certain extent for quick understanding and information finding is done.
5. **Visualizing:** Visualization of data is done for making it fast interpretable for the users. The first five steps enlist the automated activities which converts signals into information. Till now, the signals are converted into meaningful and user interpretable information. The further steps of the process will enlist non-automated activities which involve humans to interpret the information, thereby acting on them.
6. **Interpreting:** Humans interprets the information generated in earlier steps.

7. **Internalize:** In this step, the information derived after interpretation is combined with the contextual information of problem. In addition, an analysis of real underneath trends and explanatory factors for embedding the information is embedded according to cognitive understanding of the system.
8. **Revise and Recalibrate:** The new information may affect prevailing information. During this step revision and adjustment of prevailing information is done according to the new information.
9. **Verify:** This step checks the fresh information with other methods.
10. **Enrich:** During this step, the information consisting of graphs, figures, numbers, etc.—are augmented by textual justification of the information.
11. **Share and Communicate:** As we share and communicate the information within the organizational members, various discernments and outlooks are obtained.
12. **Remember:** There lies a requirement to remember some information which may not require immediate action but can be relevant in near future.
13. **Decide:** During this step, managers decide on the way to act on the information being generated.
14. **Distribute:** The decision being taken by managers on higher levels of the organization are distributed to the right persons within the organization.
15. **Anticipate on Changes:** The new information generated can be of a negative character, which requires certain change in (structural) organization. An organization should develop a positive attitude for this alteration in accordance with the newly generated information.

These fifteen steps illustrate the translation of a signal into information for the action of managers.

1.3 SOCIAL BUSINESS INTELLIGENCE (SBI)

1.3.1 THE CURRENT STATE OF SOCIAL BUSINESS INTELLIGENCE (SBI): EARLY ADOPTION

The opportunities produced on social media platforms for firms are often acknowledged by software developers. The growth of social media and the reputation of BI amongst organizations, software solutions which offers social media 'intelligence' have grown rapidly. Consequently, tools for analysis of information become extensively accessible at cost-effective prices [9], some are even offered free of cost.

1.3.2 EMERGING SOCIAL MEDIA BUSINESS INTELLIGENCE VENDORS

Companies which are based on the application of social media in their organization runs a cycle constituting of three steps; (i) monitoring, (ii) analyzing, and (iii) engaging [10, 11] that uses social media monitoring platforms. The aim of these platforms is 'listening,' for monitoring the brand(s). The collected data in this phase is constituted by mapping customer perceptions, sentiments measurement with the suggestion of the company's achievement with respect to social media.

The promising social media intelligence tools are described as follows:

1. **Radian 6:** It offers social media monitoring tools, social media engagement software and social customer relationship management (CRM) and marketing software. It offers companies with social analytics that constitutes social media metrics and sentiment analysis. It also offers firms with a dashboard for demonstration of its performance on social media. Accordingly, online discussions can be scheduled.

2. **Kapow:** It proposes solutions for the access, extraction, and enrichment of web data [10]. The application of public web data for business intelligence is demonstrated by software developer which will mention the software providing framework for social media data thereby converting it into deducible information. One of the major tools offered by Kapow is the monitoring of social media platforms.

3. **evolve24:** It will mine precedence's and convert the scores online for providing applicable intelligence to the management. Its software permits user in creation of custom dashboards for the presentation of social media metrics which have relevance with the firm. It also allows predictive modeling for the prediction of results of certain issues, since the decision, making procedure in the organization goes at par with the information [12].

4. **NetBase:** It permits users to path social media concerns connected to the interested topics. It processes numerous social media posts for extraction of structured insights which can be used by the enterprises for quick discovery of market needs and trends, quantify market acuity regarding products, services, and companies [13].

1.3.3   *INTELLIGENCE PROVIDED BY SOCIAL MEDIA MONITORING TOOLS*

Social media tools, which comprises of monitoring, analysis or intelligence tools, contribute to the performance of firms on social media in various ways. The innovation of social media and its applications on business intelligence is still unexplored. Therefore, a very few scientific literature is available in this area. Most of these documents denote the same variables for measurement of their self-discovered attributes. Prevailing attributes of social media and intelligence which are provided by the social media monitoring tools are enlisted herewith.

1. **Volume of Posts:** It quantifies the amount of messages or articles concentrated on a topic that have been created on social media within a time frame. The volume of created social media posts which contains firm's name (or product/service name) demonstrates the limit up to which a company's topic is discussed in social media. It can change from day to day or even from hour to hour.

2. **Engagement:** It represents the extent up to which a user is involved with the brand. Generally, this engagement is measured through the quantity of likes, followers, shares, retweets, etc. Doeland [14] differentiates engagement metrics into *distribution* metrics and *interaction* metrics. Distribution metrics elaborates the extent of visibility of an organization to social media public, while interaction metrics describe the limit of public engagement with the brand.

3. **Sentiment:** Most of the software tool offers sentiment analysis, which is a measurement representing the attitude of the content produced by the social media users. Commonly, social media posts are categorized by positive, neutral or negative by linguistic algorithms. These algorithms 'simply' mines the text of each post connected to the organization and connects words and phrases like

'great,' 'wow,' 'good,' ':-),' 'super,' etc. with a positive attitude. Posts containing words like 'bad,' 'dumb,' 'worthless,' etc. are categorized as negative posts.

4. **Geography:** Upon registering on social media platform, a person needs to produce some personal information which includes his residence. Although authentic information is not guaranteed but social media monitoring tools makes use of this information for determination of the location where the posts has been made. In addition, cellular devices can use a GPS component for providing the social media post with more precise geographic information.

5. **Topic and Theme Detection:** Social media monitoring tools has provision in providing details in the primary topics and themes consisting of dataset in relation with the firm. Topic and theme detection permits firms to accumulate understanding of frequent discussed topics which embodies the social media posts in relation with the firm.

6. **Influencer Ranking:** Nearly all social media platforms-and particularly social networking sites-supply the possibility to follow other users. As an outcome messages that is shaped by people with numerous followers comes within scope of many other users. Social media monitoring tools will aware the numerous followers of the people who posted a message that contains firm names.

7. **Channel Distribution:** For the assessment of the social media platforms firms which are discussed by social media monitoring tools supplies awareness on the circulation of posts connected with the firm among dissimilar platforms. Thereafter firms can undertake a decision focusing those platforms which are under the topic of discussion.

1.3.4   REQUIREMENTS FORMULATION

After a study on the social media monitoring tools which are grounded on the knowledge we have gained while performance of content analysis (CA) on the social media messages linked to distinct firms, the following necessities for business intelligence methodology have been generated.

1.3.4.1 DESCRIPTION OF REQUIREMENTS

1. **Having Access to Social Media Platforms:** Firms should have access to different platforms where messages are produced for generating intelligent information from social media data.

2. **Identification of Social Media Platforms Under Discussion:** A survey on platform distribution will cater firms with an idea about which social media platforms should focus, engage or advertise. Although firms, the matter might be discussed on numerous platforms but all of them need not be monitored separately. These tools offer monitoring and engagement of different social media platforms in *one* dashboard.

3. **Identification of the Volume of Social Media Messages Relevant to the Firm:** Consistent monitoring of the social media messages focusing on the firm permits the occurrence of sudden deviations, demonstrating that "something going on," and requires a special attention from the management.

4. **Removal of Spam from Social Media Messages which Might be Related to the Firm:** Abundant social media messages containing firms in their posts, doesn't have any relation with the firm which is a drawback as they will not detect all firm related messages uses generic names in posts. Spam messages should necessarily be removed from dataset since they are valueless.

5. **Anonymization of Personal Data:** Because of the new regulation by European Commission processing personal data, firms are disallowed from processing which allows one to retrace a natural person from that data. To work according to the new Regulation, firm plans in collection and processing the social media data for its anonymization.

6. **Identification of People Who Discusses the Firm on Social Media:** Although Kaplan and Haenlein [15] demonstrated that the usage of social media is diversified according to the users' age, it is reasonable in determination of people who discusses about the firm on social media.

7. **Identification of Subjects of the Social Media Messages Connected to the Firm:** It is also precious for a firm to have an idea about the matter which social media users confer regarding the company.

8. **Determination of Information that Contains Additional Firms Value in Social Media Messages:** A CA of a set of social media messages in relation with firms reveals that there are some messages which involve the firm's name, but doesn't enclose any information of any value which are referred as *undefined* posts.

9. **(Automatically) Classification of Social Media Messages Related to Categorization of Firms:** Due to the unstructured character, social media posts of messages needs to be pre-processed before commencement of analysis. Classification of the messages in categorized manners like subjects, languages, men, and women, many or less followers, etc. permits a firm in structural analysis of messages and derivation of the particular information which the firm is interested in.

   The unstructured nature and the large amount of messages which are generated in context with some firms makes the social media data termed as "big data." So there lies a requirement that the categorization method of the social media messages runs automatically by automatic classifiers.

10. **Form a Relation of (Categories of) Subjects of the Social Media Messages to the Firm's Key-Performance Indicators:** There lies a possibility of the classification of social media posts into categories in relation to KPIs. The subjects of the social media posts serve the basis in assignment of a certain social media message to a certain key-performance indicator.

11. **Determination of Firm's Social Reputation:** A SBI methodology should lead to the determination of the firm's influence on social media. The volume of messages in relation with the firm does not have any value if there is no awareness in the nature of these messages. Social media monitoring tools offer the possibility in determination of the sentiment of a social media post.

12. **Determination of Social Reputation to the Firm's Product(s)/ Service(s):** Along with the firm's social reputation, it might be interested in the reputation of a product or service which it supplies. Therefore, sentiment analysis is a requirement for these posts. The social reputation of products-calculated by the percentage of positive posts in relation with that product or service-reveals a relation with the sales or quantity of returns of that product.

13. **Determination of Relations between Social Media Metrics and the Firm's (Social) Key-Performance Indicators:** Preliminary objective business intelligence is identification of activities of a

firm deliverable value. For determination of which social media metrics forms a relation to the firm, the SBI procedure necessarily contains a stage where the affinity amongst social media metrics and the firm's KPIs are determined.

14. **Updating the Status of the Social Media Metrics and the Values of the KPIs Constantly:** For the development of real-time business intelligence, the system must automatically monitor the social media metrics. The firm gets an affinity on the values of the social media metrics and the values of the KPIs.

15. **Presentation of Slope of the Relations between Social Media Metrics and KPIs on a Time Chart:** This necessity assures that the slope of the values is demonstrated in such a way so that deviations over time are recognizable easily.

16. **Interpretation of the Gained Intelligence and Cater the Firm's Development:** The derived intelligence can present the relation of social media metrics and KPIs and providing a vision of the external stakeholders' perception on the firm, we should position this intelligence within the developments of firm.

17. **Assignment of Gained Intelligence within the Right Persons in a Firm:** Whenever situation arises, some KPIs are affected social media metrics, and are under performing, the acquired intelligence should be disseminated to the accountable departments within the firm.

18. **Allowing a Firm in Engagement on Social Media Platforms:** A SBI methodology should permit firms in engaging with the users on social media through social media discussions.

19. **Regularly Updating the Search Terms in Anticipation of Changes:** A fluent SBI method starts with search terms in relation with the firm. Since a firm always develops, it will initiate new products, services, and employees will come and go. As a result, the search terms should be updated when there are events which influence the required search terms.

1.4   SOCIAL BUSINESS INTELLIGENCE (SBI) PROCEDURE

A prototype for SBI procedure has been developed whose aggregate overview is presented in Figure 1.2. This methodology consists of seven main activities which are related to each other.

1.4.1 STRATEGIC MAPPING OF KPIS

Firms deduce KPIs from their plans. The KPIs which a firm eventually establishes are to be calculated by CA. Some KPIs are best determined by internal systems, whereas others are perfectly determined by social media. From the list of KPIs which a firm uses, a choice can be made on indicators which can be calculated in terms of social media.

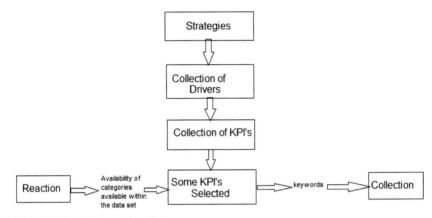

FIGURE 1.2   SBI procedure [16].

These KPIs commences the SBI procedure, since for them the social media data is to be gathered and evaluated. The KPIs preferred to be calculated by social data for determination of categories which are to be analyzed including the subjects of social media messages-and accordingly keywords which are to be worn for the collection process.

1.4.1.1 COLLECTING

After the selection process is over by KPI measurement, the data is to be *collected*. The step in Figure 1.3 illustrates Keywords in relation with the firm, its products/services, and the selected KPIs worn to "listen" to numerous social media channels where the firm could be specified. The CA affirms that it differentiates per firm where social media platform are discussed. So, the first step involves social media platforms constituting the purpose of platforms where the firm is discussed. Search queries in relation with firms will lead to unstructured data from different social media podiums.

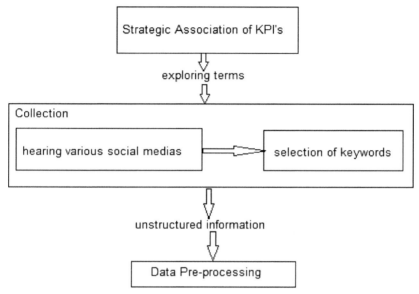

FIGURE 1.3　Blueprint of SBI procedure (collecting) [16].

1.4.1.2　DATA PRE-PROCESSING

The third step abides *pre-processing* of the composed facts. As compared with 'regular' BI statistics, social media data are formless, obtained after numerous podiums, contains junk with individual statistics, consequently requiring being pre-processed. Figure 1.4 demonstrates this procedure. The collected data abides social media messages obtained after different foundations in dissimilar layouts, for instance, CSV, JSON, XML, etc. Every statistical foundation can apply its individual assembly on social media communications, whereas every platform may not include the same importance in attributes with others. Every social media post should be parsed, structured-into a single data format. With the scraping method, numerous search queries produce multiple messages which might be present in the database. As a result, solitary social media communications which do not happen in the counter must be supplementary. This ultimate phase in the *data pre-processing* stage abides the removal of spam. As data *pre-processing* concludes, facts becomes organized, cleaned besides prepared for categorization.

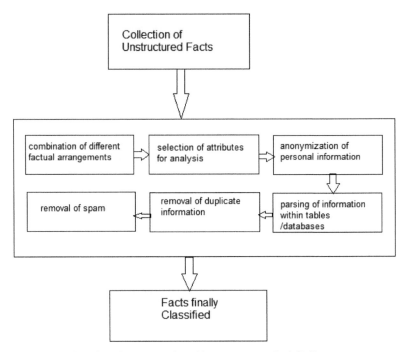

FIGURE 1.4   Blueprint of SBI procedure (data pre-processing) [16].

1.4.1.3 CATEGORIZING

The motive of this phase lays the division of the communications hooked on gathered groupings on which a company is concerned. The different conditions at which the social media stakes are characterized might differ. Figure 1.5 illustrates the third phase of the SBI technique. We can make a decision analysis of people who creates the messages, and group these people in. A labeling is done on the four classifications of individuals. Encouragers through fewer supporters although speaking positively around the company or its yields. Protestors through fewer supporters besides writing negatively around the company. Individuals through numerous supporters who express positively around a company are defined as organizers, although individuals consisting of numerous support lettering negatively labeled saboteurs. A scrutiny of these individuals supplies company through intellect of the influence of individuals who writes around company, which might custom the beginning argument of a social media appointment procedure.

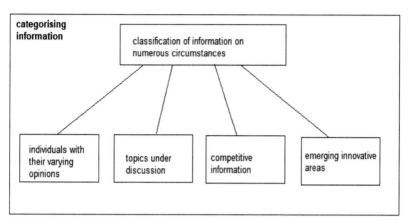

FIGURE 1.5 Blueprint of SBI procedure (categorizing) [16].

Additional aspect by which social media communications might be categorized remains according to their themes. The subjects which characterized our dataset relates to communal image, purchaser associations, employment, merchandise besides service quality, product, and service innovation, professionals' opinions, etc. With the classification of posts into categories based on subjects, there lies a possibility in linking the volume of messages in relation to certain subjects of the companies' analogous KPIs. There are many indifferent categories for the categorization of social media communications, for intercommunications with company's KPIs to social media facts, classification must be done on the messages according to subjects [16].

1.4.1.4   ANALYZING

After the social media data is structured and cleaned, there is commencement of the *analysis* of data. During this step, interpretation remains through statistics to information. This phase is elaborated in Figure 1.6. Based on these specific subjects an analysis can be done on various facts and associations. It might be reasonable in at slightest map the volume of conversation-before quantity of social media notes in relation with company-contrary to dissimilar social media frequencies for determinations of extent of conversations in relation with firm. Adjoining the extent a category has been established in previous step where entire messages in relation with certain product or product characteristics are grouped, and then there lies a possibility in determination of contempt of the public on product with application of sentiment analysis on these data. This analysis supplies the firm with view on the products or product characteristics that should be upgraded. Moreover, an analogous investigation proceeding contenders' social media statistics resolve demonstrate the company's situation in comparison with the contenders besides their yields.

The intellect being obtained during the analysis phase reveals some social media metrics remains to perform less besides these social media parameters provokes the KPIs of a corporation. Subsequently, a firm might issue necessary activities for improvement of these parameters.

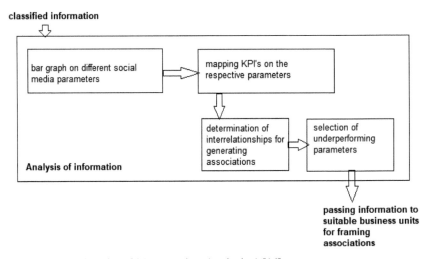

FIGURE 1.6   Blueprint of SBI procedure (analyzing) [16].

1.4.1.5   *MAPPING INSIGHTS TO BUSINESS UNITS*

KPIs remain connected toward various subdivisions within a stable, besides the executives of these subdivisions can provide a clarification of low performing parameter and thereby initiate certain methods for improvement of KPIs.

1.4.1.6 REACTING

The concluding stage in SBI procedure consists of the implementation of action plans for improvement of less performance KPIs in expressions of social media. A firm can undertake a decision for reviewing products (features) on the basis of complaints and suggestions that SBI procedure has supplied [16].

1.5 REAL-TIME SOCIAL BUSINESS INTELLIGENCE (SBI)

Social media data is made precious because of its real-time and direct availability. During analysis, a scrutiny for newly generated social media messages is implemented every day. Upon the diversion of social media activity of a firm from steady-state, information should be passed.

1.5.1 BUSINESS INTELLIGENCE FOR DECISION SUPPORT

SBI defines generation, publication, and share of traditional commercial analytics accounts besides consoles through conclusive operators of Cloud technologies. Enterprise 2.0 apparatuses besides follows through commercial intelligence outputs are used by social BI for undertaking any collective decisions. Social BI permits cooperative expansion of post-user-driven analytics between business forecasters besides data mining (DM) specialists which removes earlier hindrances to self-provision BI with implications of outdated analytics submissions [17].

Social BI interprets in supplying business intelligence build upon social networks data. The report and visualization build upon social media symbolizes on real experiences of people. Extracting data from various social media to generate comprehended reports facilitates company in deciding subsequent actions. The dashboard, visualization, report confining social media provides companies in obtaining adequate feedbacks and take actions [17, 18].

Judgments undertaken by BI governance committee is held accountable in alignment of commercial calculated creativities besides methodologies through BI presentations, speculation, with practice [19, 20].

Facts symbolize precious advantages which is crucial element for generating a decision [21]. Since past, a BI application permits users for obtaining knowledge through company-internal data by means of numerous technologies. This 'explosion of data,' with special mention of dimensions, variability, and rapidity of the prevalent facts, leads in determination of a shift to big facts besides implicates new mode of BI [22].

1.5.2 SOCIAL BUSINESS PERSPECTIVE

Presently individuals are convenient in sharing data and outlooks by social surroundings. Rendering to [23] corporations may establish a "social intelligence" build upon information. Thoughts dispersed by social-networking through employees, customers, and even external players. A positivity in corporate culture, stimulating imagination and innovation uplifts communities in virtual like "discussion areas" by simulation of organizational members acting beneficially. The settled "discussion areas" stretches around company's arena of movement besides attention which are continued by a large area of social tools, systems, and technologies. A gathering of social media matter shall be evaluated besides prepared for gathering precious knowledge which enhances outlook of company. Subsequently, conclusions will be enhanced. The method of collection of social facts, its scrutiny for undertaking improved conclusions is defined as SBI [24, 25], contrarily BI method is enhanced with "social intelligence." Merging social facts through outdated interior besides exterior facts establishes a distinct aspect which is imported into the decision-making scheme.

The advantages of social media snooping (Figure 1.7), appropriate apparatuses, schemes besides machineries which will facilitate in separation of signal from noise has been disclosed by "marketing executives" [26].

Information centered in consumers, their comforts besides acquiring conclusions is converted hooked on information coupled with the administrative understanding disreputable and build dominion conclusions. Physique-in investigative competences, key performance indicators (KPIs) aiming in evaluation of data are accessible to decision makers. Executives are deceptive on worth created with social information within areas comparable to resource restraint organization, merchandise expansion, character supervision, and danger administration [24].

FIGURE 1.7   Social snooping. Integration of social facts inside the BI method [26].

Accordingly, communal schmoosing has involved association, administrators, specialists, to become active in creation of social data inside the establishment [25, 26]. Communal statistics has opened a new approach in taking decisions for the judgment fabricators.

1.5.3   INTEGRATING THE SOCIAL VIEW INTO A CLASSICAL BI SCHEMA

Management of social data is demanding, therefore an acceptable data warehouse architecture is required. The idea initiated a social structure aimed at accustoming some conventional BI schema towards a communal pathway. By means of SBI procedures and magnitudes, reengineering will be satisfied [27]:

- The preliminary detail counter will be translated hooked on measurements-Old_Fact_Now_Dimension; consequently, the procedures of conventional BI prototypical will be converted to elemental characteristics-Am_1, AM_2, …, AM_m; in contrary with the translation the communal standpoint of the entire method might be explained as an add-one capability of traditional plan;
- Determined difficulties along-with the guidance in providing solution to these problems becomes elemental characteristics of binary supplementary aspects-Difficulties, Movements;
- The personalities recognized as performers around fundamental societal community is modeled as a differentiate measurement-Characterizations; modeling of measurement on the basis of individuals' character surrounded by the communal surroundings besides their functionalities for the corporation;

- Establishment of the scope of SBI data model-MS_1, MS_2, …, MS_n; which drives the aggregated in comparison through newly evolved magnitudes-Old_Fact_Now_Dimension, Problems, Movements, and Descriptions.

Negligible alterations of the proposed SBI schema may be considered in tangible enactments, e.g., establishing Movements besides Difficulties within a dimensional pyramid with implanting Time as a supplementary dimension.

1.6 SOCIAL NETWORK FOR DECISION SUPPORT

Group and personal decision-making support, varies by high lightening the fact of group decision-making support in organization cannot happen without crucial adaptations. Those alterations grow from the necessity of integration of multiple teams around highly collective and composite (though intertwined) group decision procedure, creating a necessity for the management of dependencies amongst individuals, measures, administrative components, and artifacts. Alongside, three prime kinds of addictions might be studied: movement, distribution, and variation. The movement appears once a movement needs an outcome of an alternative. Distribution of enslavements appears once various happenings claims identical source (individuals, machineries, planetary, etc.). Lastly, version appears since the requirement of an appropriate fitting amongst happenings of organization.

With the expansion of complications within an organization there, lies a necessity in solemnize besides standardizing them. Instead of implanting a definite explanation, officially termed procedures should act as a backbone in providing a definite assembly to the specific with communal performance [28].

It is effortless in considering that decision-making can serve to be advantageous on existence of framework of organization till becoming definite and enforced procedure. Feldman and Pentland [29] elaborated that, inside administrative boundaries, two categories/models of associated acts exist: the ostensive and the performative. These organizational structures are a powerful source in explanation of organizations flexibility, behavior, change, and, thereby powerful information source in estimation of opportunities and constraints to undertake decisions in organization.

The apparent models obtain an accurate description about the ways of operation and execution of methodologies supported with examples, which uses tools like commercial procedure modeling, symbolization, and UML illustrations (specifically, performance besides communication), aimed at description of workflow procedure. These explanations generally erected into manuals for working procedures which apparently explore in appointing a precise besides comprehensive method aimed at performing duties along-with otherwise procedure.

Performative customs focuses explanation of ways of duties or procedures carried out importing adjacent to the grounded theory research methodology [30] which manuscripts, evidences, furthermore cautiously analyzes the behavior of individuals, establishing a description or tolerance for the procedures and arrangements influenced by them.

While a focus is made on conventional supervisory procedure realization becomes easier that it efforts huge quantity of undertaking-arrangement within groups, so the performances are enclosed with an ostensive methodology since the judgment procedure exploits a mechanized nature commencing three-phase model of intellect, plan, with alternatives [31]. Here the procedure conceivable originates starting contradictory towards convergent condition by a distinct succession, with pursue of iteration procedure [32]: (a) quandary investigation and description occurring within astuteness stage; (b) disagreement sustained by generating alternatives with its evaluation by collection, the meeting procedure originates; (c) throughout proposed stage, available outcomes for the problems be achieved (disagreement) pursued through joining of similar thoughts with removal of unnecessary or immaterial thoughts (meeting); (d) selection comprises of contradictory assessment of prevalent thoughts through convergence choice. For expanding the usage of the earlier "traditional" structure with amalgamation of organizational decision-making, the organizations should also revolve around ways the procedures are really accomplished, alternatively only regard the way to be carried on, if an existent interest to build an equal managerial collection conclusion sustain organization. The communal arrangement model [33] being used to organize the communal arrangement intended for judgmental sustainance elaborates mapping which essentially takes place anywhere the space (position) is regarded within a broad and relatively effective way. Since connections of facts, estimations, conviction, with vigor are considered indispensable in organizational methodology, this model places the important players for possible reasons and interruptions while communicating, or alliance, the subgroups build on temporary ground, and so forth.

This model is moreover worn as methodology or investigative instrument [34] whiles the study of relationships and interactions amongst unlike actors. This procedure is commonly used in virtual and computational environments [35], has gained an importance in the last decade. Alongside the prevalent tools for helping this kind of amalgamation are generally made for recording effortlessly and mechanically evidence, records, and obtain large amount of data for analysis [36], to make it adaptable in grounding an effectual assemblage judgmental sustainance within an organization. Figure 1.8 combines earlier deliberations, with viewing a complication augmentation within assemblage judgmental sustainance, with advancement from personal to networked group decisional aids. The organizational network elaboration demonstrates that ostensive procedures in decision-making can grant themselves unequal in some activities by determination of versification within the group adaptation. Here the usage of formal methods as structure to decide about the following rigid method. The benefit of performing replicas, for instance social network model, is valuation of genuine procedures with bestow on reassessment of formal procedures, promotion of organizational flexibility and efficient assistance for combined organizational group decision support system.

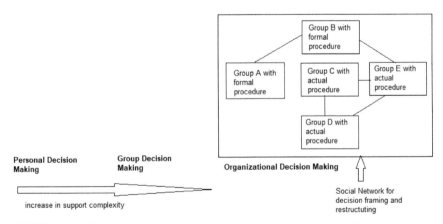

FIGURE 1.8 Usage of social network model for managerial assemblage managerial sustainance [36].

1.7 CONCLUSION

Social media forms budding area within educational arena. Prevalent investigation within areas of social media is mostly objected in promotional exertions otherwise alternative work within the outlooks of the firm else should demonstrate into the outer arenas. The chapter concentrates on the acceptance of information from the organization perspective, focusing on evolution of facts commencing social medias for undertaking any decisions. The area which this chapter investigates lies on the application of communal facts for taking any decisions within organization, thus providing with huge prospects in the world of business intelligence. In this respect, the chapter is distinguished from other areas of research.

KEYWORDS

- **business intelligence**
- **decision making**
- **perceptions**
- **real-time social business intelligence**
- **social business intelligence**
- **social network**

REFERENCES

1. Mitra, A., Satapathy, S. R., & Paul, S., (2013). *Clustering in Social Network Using Covering-Based Rough Set.* Paper presented at IEEE-Int. Conf. (IACC 2013) Ghaziabad, India.
2. Mitra, A., Paul, S., Panda, S., & Padhi, P., (2016). *A Study on the Representation of the Various models Dynamic Social Networks.* Paper presented at Elsevier-Int. Conf. (ICCCV 2016), Mumbai, India.
3. Elbashir, M. Z., Collier, P. A., & Davern, M. J., (2008). Measuring the effects of business intelligence systems: The relationship between business process and organizational performance. *International Journal of Accounting Information Systems*, *9*(3), 135–153.
4. Van Beek, D., (2006). The Intelligent Organization. Performance improvement and organizational development with business intelligence.

5. Rouibah, K., & Ould-ali, S., (2002). Puzzle: A concept and prototype for linking business intelligence to business strategy. *The Journal of Strategic Information Systems, 11*(2), 133–152.

6. Smith, R., (2006). *Key Performance Indicators.* The basics of maintenance and reliability - Chapter 6.

7. Tsai, Y., & Cheng, Y., (2012). Analyzing key performance indicators (KPIs) for e-commerce and internet marketing of elderly products: A review. *Archives of Gerontology and Geriatrics, 55*(1), 126–132.

8. Shahin, A., & Ali, M. M., (2007). Prioritization of key performance indicators: An integration of analytical hierarchy process and goal setting. *International Journal of Productivity and Performance Management, 56,* 226–240.

9. Bughin, J., Chui, M., & Manyika, J., (2010). *Clouds, Big Data, and Smart Assets: Ten Tech-Enabled Business Trends to Watch.* McKinsey & Co. Quarterly (White Paper).

10. Kapow, (2009). *Five Key Lessons for Converting Social Media Data into Business Intelligence.* Kapow Software (White Paper).

11. Bryant, L., (2011). *Moving from Social Media Monitoring to Social Business Intelligence.* Head shift, Dachis Group. Retrieved from: http://www.headshift.com/our-blog/2011/06/15/from-social-media-monitoring-to-social-business-intelligence/ (accessed on 25 February 2020).

12. Evolve24, (2012). *Website.* Retrieved from: http://www.evolve24.com/.

13. Netbase, (2012). Website. [Online] http://www.netbase.com/our-network/partners/sapsocial-media-analytics/ (accessed on 25 February 2020).

14. Doeland, D., (2012). *Social Media and Internet Key Performance Indicators.* Retrieved from: http://ddmca.nl/diensten/social-media-and-internet-monitoring/social-media-key-performance-indicators/ (accessed on 25 February 2020).

15. Kaplan, A. M., & Haenlein, M., (2010). Users of the world, unite! The challenges and opportunities of social media. *Business Horizons, 53*(1), 59–68.

16. Heijnen, J., (2012). "Social Business Intelligence." *Master Thesis.* Technische Universiteit delft.

17. Forte Consultancy Group, (2014). *Social BI for Intelligent Enterprise 2.0.*

18. Group Charger, (2011). *Social Business Intelligence.*

19. Larson, D., & Matney, D., (2007). The four Components of BI Governance. [Online] http://www.bibestpractices.com/view/4681 (accessed on 25 February 2020).

20. Muntean, M., Muntean, C., & Cabău, L., (2013). Evaluating business intelligence initiatives with respect to Bi governance. *Proceedings of the IE2013 International Conference.*

21. Kaplan, R., & Norton, D., (1996). *Translating Strategy into Action.* The Balanced Scorecard, Haward Business School Press Boston.

22. Blomme, J., (2012). *The New Normal in Business Intelligence.* http://www.slideshare.net/johblom/the-new-normal-in-business-intelligence (accessed on 25 February 2020).

23. Harrysson, M., Metayer, E., & Sarrazin, H., (2012). *How 'Social Intelligence' Can Guide Decisions.* McKinsey Quarterly. http://www.mckinsey.com/insights/high_tech_telecoms_internet/how_social_intelligence_can_guide_decisions (accessed on 25 February 2020).

24. Palmer, D., Mahidhar, V., Galizia, T., & Sharma, V., (2013). *Reengineering Business Intelligence.* Amplify Social Signals, Business Trends, Deloitte University Press.

25. Kobielus, J., (2007). *Business Intelligence Gets Collaborative.* Network World. http://www.networkworld.com/columnists/2007/011507kobielus.html?page=1 (accessed on 25 February 2020).
26. Jain, U., (2012). Collaborative BI: Wisdom of the crowds, software magazine. The Software Decision Journal. [Online] www.softwaremag.com/content/ContentCT.asp?P=3377 (accessed on 25 February 2020).
27. Muntean, M., & Cabău, L., (2011). *Business Intelligence Approach in a Business Performance Context.* Austrian Computer Society, Band 280.
28. Poltrock, S., & Handel, M., (2009). Modeling collaborative behavior: Foundations for collaboration technologies. In: *Proceedings of the 42nd Annual Hawaii International Conference on System Sciences.* Computer Society Press, Big Island, Hawaii, USA.
29. Feldman, M. S., & Pentland, B. T., (2003). Reconceptualizing organizational routines as a source of flexibility and change. *Administrative Science Quarterly, 48*(1), 94–118.
30. Glaser, B. G., & Strauss, A. L., (1967). *The Discovery of Grounded Theory: Strategies for Qualitative Research.* Aldine, Chicago, Ill, USA.
31. Simon, H., (1977). *The New Science of Management Decision, Prentice-Hall.* Englewoods Cliffs, NJ, USA.
32. Pendergast, M., & Hayne, S., (1999). "Groupware and social networks: Will life ever be the same again?" *Information and Software Technology, 41*(6), 311–318.
33. Cross, R. L., & Parker, A., (2004). *The Hidden Power of Social Networks: Understanding How Work Really Gets Done in Organizations.* Harvard Business School Press, Boston, Mass, USA.
34. Wasserman, S., & Faust, K., (1994). *Social Network Analysis: Methods and Applications.* Cambridge University Press, Cambridge, Mass, USA.
35. Harrer, A., Moskaliuk, J., Kimmerle, J., & Cress, U., (2008). Visualizing wiki-supported knowledge building: Co-evolution of individual and collective knowledge. *Proceedings of the 4th International Symposium on Wikis (WikiSym '08).* Porto, Portugal.
36. Nooy, W. D., Mrvar, A., & Batagelj, V., (2005). *Exploratory Social Network Analysis with Pajek.* Cambridge University Press, Cambridge, Mass, USA.
37. Francisco, A., & Joao, P. C., (2012). Integrating decision support and social networks, *Advances in Human-Computer Interaction.* Hindawi Publishing Corporation. doi: 10.1155/2012/574276.

CHAPTER 2

Critical Concepts and Techniques for Information Retrieval System

MOHD SHAHID HUSAIN

Assistant Professor, Information Technology Department,
College of Applied Sciences, Ibri, Ministry of Higher Education, Oman,
Tel.: 968-94714261, E-mails: siddiquisahil@gmail.com,
mshahid.ibr@cas.edu.om

ABSTRACT

This is the era of information technology. Today the most important thing is to get the right information in right time. The users of internet and other communication technologies are increasing exponentially. Hence more and more data repositories are now being made available online information retrieval systems or search engines are used to access electronic information available on the internet. These information retrieval systems depend on the available tools and techniques for efficient retrieval of information content in response to the user query needs [1]. Most of the work done by researchers in the field of information retrieval Systems is text oriented and monolingual. These IR systems are basically focused on English and other European languages. Due to the exponential growth in the use of internet, now we have a lot of information in the form of images, videos' and audio form. Hence there is a need of information retrieval systems which can retrieve multimedia information as per the user's requirements. In this chapter we will deal with the basic concepts of information retrieval systems, there needs, models of information retrieval systems and other related concepts like stemming, indexing, etc. the chapter will help scholars and young professionals to get critical information required for developing IR systems.

2.1   INTRODUCTION

In this era of information technology, more and more data is now being made available on online data repositories. Almost every information one need is now available on internet. The term "information retrieval" was coined by Calvin Mooers in 1950 [2]. In simplest form, we can define information retrieval as: finding material (usually documents) of an unstructured nature that is relevant to the user from within huge collections available on internet. A relevant document contains the information that a person was looking for when they submitted a query to the search engine, i.e., it satisfies the information need of a user.

Theoretically there is no constraint on the type and structure of the information items to be stored and retrieved with the information retrieval (IR) system. Until recently information retrieval systems were limited to searching textural information. Nowadays multimedia indexing and retrieval techniques are being developed to access image, video and sound database without text descriptions.

Also, English and European languages basically dominated the web since its inception. However, now the web is getting multi-lingual [3]. Hence automatic information processing and retrieval is become an urgent requirement.

2.1.1   OVERVIEW

The use of digital technologies and growth in technological developments for storing, manipulating and accessing of information has led to development of valuable information repositories on internet (Figure 2.1).

The rapid growth of electronic data has attracted the attention in the research and industry communities for efficient methods for indexing, analysis and retrieval of information from these large number of data repositories having wide range of data for a vast domain of applications.

2.1.2   DRIVING FORCES BEHIND THE DEVELOPMENT OF AN EFFECTIVE IR SYSTEM

As the internet users around the world are growing day by day, the quantity of the electronic data available on the internet is also increasing. Now we have huge amount of information available on internet, not only in English or some other European languages but also in various regional languages. To access this valuable information available on internet, it becomes essential to develop some retrieval systems/search engines supporting these languages. Especially in the case of Asian Languages or in particular Indian languages, lots of data is being available on the internet in various regional languages like Hindi, Urdu, and Telugu, etc. Accessibility to this information is very limited because there are very limited resources to support efficient retrieval of information for these languages.

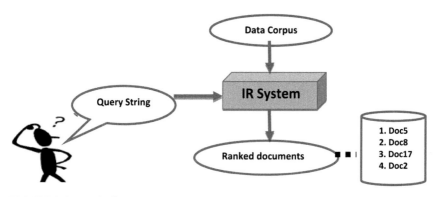

FIGURE 2.1   Typical IR system.

An Information Retrieval system is the system which retrieves some ranked documents in response to the user query. An efficient Information retrieval system satisfies the user's information need. However, only a limited proportion of the world population uses English as their primary language to express their information need. So there is a need of some effective tools to support the information retrieval in other languages [4].

2.1.3   INTRODUCTION TO IR

Information retrieval is the sub domain of text mining and natural language processing. This is the science in which the software system retrieves the relevant documents or the information in response to the user query need [1].

A user gives a query according to his information needs and the IR system retrieves the most relevant documents containing that information.

The most widely used IR application is web search engines, which helps users to get information from the internet according to his needs.

The Information Retrieval system match the given user quires with the data corpus available and rank the documents on the basis of the relevance with the user need. Then the IR system returns the top ranked documents containing relevant information to the user query.

Figure 2.2 shows the architecture of the mono-lingual information retrieval system. This is a basic architecture and may be used for the development of any mono-lingual information retrieval system.

Information retrieval systems consist of different modules like query expansion, document indexer, user interface, document ranking, etc. as in the Figure 2.2.

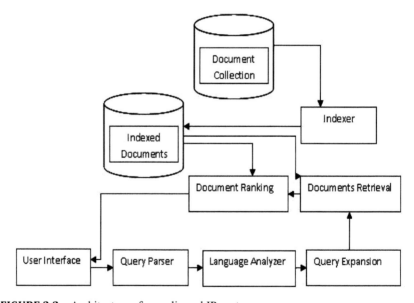

FIGURE 2.2   Architecture of monolingual IR system.

2.1.4   IR PROCESS

The objective of an IR system is to analyze the user's query and retrieve the most relevant documents in a ranked manner which satisfies the information need of the user. To retrieve the relevant information on the basis of user query IR system performs the following process (Figure 2.3):

- The user gives a query statement.
- The Information Retrieval system converts the given query phrase and the documents presented in the corpus in a standard format.
- A matching function is used by the system to match the query word/ phrase with the documents available in the data corpus.
- The system then ranks these documents on the basis of the relevance with the query.
- Finally the system retrieves the top ranked documents and presents them to the user.

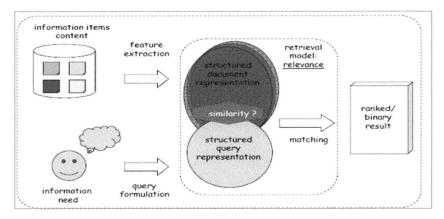

FIGURE 2.3   IR process.

There are different techniques used for transforming the query word/ phrase and the documents in the corpus in a standard format like stemming/ morphological analysis, Stop word removal, indexing, etc.

Similarly there are different approaches or methods one can incorporate for matching query word/phrase with the documents for relevancy like cosine similarity, Euclidean distance, etc. There are various factors which affect the efficiency of any Information Retrieval system like term weighting schemes, document indexing strategies and the IR model used to build the Information Retrieval system.

2.1.5   DATABASES VS. IR SYSTEMS

We are using Databases (RDBMS) for a long time for storing, maintain and accessing information. Information Retrieval systems are also used for storing, managing data and retrieval of information in response to the user query. However there are some major differences between the two systems like:

2.1.5.1   TYPES OF CONTENT

In DBMS we have structured data having predefined semantics. Whereas in IR system we have mostly unstructured data.

2.1.5.2   QUERY STRUCTURE

In DBMS we have clearly defined queries based on some formal structure. Whereas in IR system generally user expresses his/her query in natural languages having free text. This creates problems like vagueness, imprecision and incompleteness.

2.1.5.3   SYSTEM RESPONSE

In DBMS user gets the exact information as per the query in formal sense. However in IR system user didn't get the exact information instead the system returns a ranked list of documents where the user can get the required information. The documents, in the list provided by the IR system, may sometimes satisfies the user's need and sometimes not.

2.2   RELATED CONCEPTS

There are various concepts like morphological analysis, stemming, stop word removal, indexing, etc. which are essential to understand to know the IR process. In this section these concepts will be discussed.

2.2.1   MORPHOLOGICAL ANALYSIS

In linguistic terms, analysis of morphology means understanding the internal structure of words. Morpheme is the smallest unit of a word having a linguistic meaning and sound. So in morphological analysis, we analyzes the structure of words and parts of words, such as stems, root words, prefixes, and suffixes and then each word is break down into its smallest basic unit having some lexical meaning for example the word "computer" will be converted into "compute."

2.2.2 STEMMER

The most basic component of any Information Retrieval system is Stemmers. Among all the morphological systems, stemmers are the simplest system.

Stemming is the process of getting the root of an inflected word by removing prefix, infix or suffix. For example, the words player, play, and playing all are reduces to the root word play by using any stemming algorithm.

The main difference between the morphological analysis and stemming is that in morphological analysis we get the base word having some lexical meaning but in stemming the root word may or may not have any dictionary meaning. Stemming is simple as we don't need any lexicon or linguistic knowledge to process the document.

In Information Retrieval systems we use stemming as it reduces the complexity of morphological analyzer and helps in increasing the efficiency of an IR system—more specifically, to increase recall to a large degree.

Looking at various techniques, Stemming approaches can generally be categorized into rule based or statistical methods. Rule based methods may require cyclical application of rules. Look-up tables of stems and/or affixes are used for defining the stripping rules and the efficiency can be increased by maintaining a dictionary. Statistical stemmers are dependent on corpus size, and their performance is influenced by morphological features of a language. Morphologically richer languages require deeper linguistic analysis for better stemming.

2.2.3 STOP WORDS

Stop words are the high frequency words in the documents which are used in a sentence for grammatical purpose. Generally these words do not convey any information about the sentiment of the document. Hence removing stop words will have almost no effect on the performance of the IR system. Indeed it helps to reduce the index size.

Some of the common words which the IR system ignores while indexing and considered them as stop words are "the," "it," "to," etc. There are different tools and API's available which can be used to remove stop words in a document.

Table 2.1 shows some sample stop words in English Language. The stop word list may differ according to the application and also it affects the performance of the IR system as sometimes phrases like "to be" can be considered as stop words however they affect the results.

TABLE 2.1   Sample List of Stop Words in English Language

A	can't	i	ought	those
about	cannot	if	our	through
above	could	in	ourselves	to
after	couldn't	into	out	too
again	did	is	over	under
all	didn't	it	own	until
am	do	its	same	up
an	does	let	she	very
any	doesn't	me	should	was
are	doing	more	so	we
as	don't	most	some	were
at	down	my	such	what
be	during	no	than	when
because	each	nor	that	where
been	for	not	the	which
before	from	of	their	while
being	had	off	them	who
below	has	on	then	who's
between	have	once	there	whom
both	him	only	these	why
but	his	or	they	with
by	how	other	this	would

2.2.4   INDEXING

We have billions of documents available from where the system retrieves the relevant documents to the user. Similarly the user query is not structured it may have different phrases or words. To represent the documents in the corpus and the user query statement indexing is done. The process of transforming document text and given query statement to some representation of it is known as indexing. Indexing reduces the size for searching and improves the performance of IR system. There are different index structures which can be used for indexing. The most commonly used data structure by Information Retrieval system is inverted index. To retrieve the information, the information Retrieval system does not use the actual text of the document and query. Instead, some representation of the documents and the query are used by the search engine. The system then matches the document representation with query representation to retrieve the relevant information.

Indexing techniques concerned with the selection of good document descriptors, such as keywords or terms, to describe information content of the documents. A good descriptor is one that helps in describing the content of the document and in discriminating the document from other documents in the collection.

The most widely used method is to represent the query and the document as a set of tokens,, i.e., index terms or keywords.

2.2.5   *TERM WEIGHTING*

To represent the importance of the terms with respect to the different documents and across the documents a term weighting matrix is created.

The factor t_f simply means the term count in a document. To represent a document term count is considered to be important because, the terms that occur more frequently represent its meaning more strongly than those occurring less frequently. The second factor considers term distribution across the document collection.

The two assumptions considered for this are:

a.   The more a document contains a given word the more that document is about a concept represented by that word.

b.   The less a term occurs in particular document in a collection, the more discriminating that term is.

There is different term weighting schemes which one can incorporate in Information Retrieval System as follows.

2.2.5.1   TERM FREQUENCY

This is a local parameter which indicates the frequency or the count of a term within a document. This parameter gives the relevance of a document with a user query term on the basis of how many times that term occurs in that particular document [1].

Mathematically it can be given as: $tf_{ij} = n_{ij}$

where n_{ij} is the frequency or the number of occurrence of term t_i in the document d_j.

Generally, in a corpus we have documents of different sizes, there are chances that some large documents may contain the considered term more frequently (i.e., have higher term frequency) compare to the some other documents of small size regardless the actual relevance of the term with the document. To normalize this effect tf is divided with the total number of frequencies of all the terms within the document. The normalized t_f can be given as:

$$tf_{i,j} = \frac{n_{i,j}}{\Sigma_k n_{k,j}}$$

where $\Sigma_k n_{k,j}$ is the sum of the frequencies of all the terms in the document d_j.

2.2.5.2   INVERSE DOCUMENT FREQUENCY

This is a global parameter which evaluates the importance of the term across the documents available in the corpus. The number of the documents in the corpus containing the considered term t is called the document frequency. To normalize, it is divided by the total number of the documents in the corpus.

Mathematically it can be given as: $df_i = n_i/n$, where n_i is the number of documents that contains term t_i and the total number of the documents in the corpus is n; and idf is the inverse of this document frequency. To further normalize it, we take log of this. So mathematically it can be given as:

$$idf_i = log\left(\frac{n}{n_i}\right)$$

2.2.5.3 DOCUMENT LENGTH

The third factor which may affect the weighting function is the length of the document.

Hence to evaluate the relevance of a document with respect to a query, the term weighting function can be represented by a triplet ABC as f = A.B.C [5], where A – tf component; B – idf component; and C – Length normalizing component.

The factor Term frequency within a document (A) may have following options as shown in Table 2.2.

TABLE 2.2   Different Options for Considering Term Frequency

n	tf = tfij	(Raw term frequency)
b	tf = 0 or 1	(binary weight)
a	$tf = 0.5 + 0.5\left(\dfrac{tf_{ij}}{\max tf \text{ in } D_j}\right)$	(Augmented term frequency)
l	tf = ln(tf$_{ij}$) + 1.0	Logarithmic term frequency

The options for the factor inverse document frequency (B) can be any of the value shown in Table 2.3.

TABLE 2.3   Different Options for Considering Inverse Document Frequency

n	Wt = tf	No conversion, i.e., idf is not considered
t	Wt = tf*idf	Idf is taken into account

The parameter document length (C) can have options as shown in Table 2.4.

TABLE 2.4   Different Options for Considering Document Length

n	W_{ij} = wt	No conversion
C	W_{ij} = wt/ sqrt(sum of (wts squared))	Normalized weight

2.2.6 SIMILARITY MEASURES

To retrieve the most relevant documents with the user information need, the IR system matches the documents available in the corpus with the given user query. To perform this process different similarity measures are used. For example Euclidean distance, cosine similarity.

2.2.6.1   COSINE SIMILARITY

This concept to check similarity between the query and the documents in the corpus, is used in Vector Space Model of IR. The documents available in the corpus and the user query are represented by the vectors in the vector space with features as axes.

The IR system rank the documents by the closeness of document vectors to the query vectors. IR system then retrieve the top ranked documents to the user.

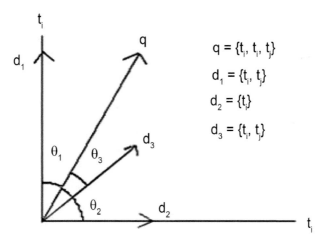

$q = \{t_i, t_i, t_j\}$

$d_1 = \{t_i, t_j\}$

$d_2 = \{t_i\}$

$d_3 = \{t_i, t_j\}$

FIGURE 2.4   A VSM model representing 3 documents and a query

Figure 2.4 shows a vector space model where axes t_i and t_j are the terms used for indexing.

The cosine similarity between the document d_j and the query vector q_k is given as:

$$\overline{d}_j \cdot \overline{q} = \left|\overline{d}_j\right|\left|\overline{q}\right|\cos\theta$$

$$sim(d_j, q_k) = \frac{(d_j, q_k)}{\left\|d_j\right\|\left\|d_k\right\|} = \frac{\sum_{i=1}^{m} w_{ij} \times w_{ik}}{\sqrt{\sum_{i=1}^{m} w_{ik}^{2}} \times \sqrt{\sum_{i=1}^{m} w_{ij}^{2}}}$$

2.2.6.2   EUCLIDEAN DISTANCE

This is the other most common method to evaluate the similarity between two or more entities. Euclidean Distance between document p and query q is given by:

$$d(p,q) = \sqrt{(p_1 - q_1)^2 + (p_2 - q_2)^2 + \cdots + (p_n - q_n)^2} = \sqrt{\sum_{i=1}^{n}(p_i - q_i)^2}$$

where n is the number of features.

2.3   INFORMATION RETRIEVAL MODELS

The IR models can be distinguished by the way how they represent the documents and query statements, how the system matches the query with the documents in the corpus to find out the related one and how the system ranks these documents. An IR model defines the following aspects of retrieval procedure of a search engine:

a.   How the documents in the collection and user's queries are transformed.
b.   How system identifies the relevancy of the documents based on the query word/phrase given by the user.
c.   How system ranks the retrieved documents based on the relevancy.

An Information Retrieval System comprises of the following components:

a.   A model for representing documents and query statement and
b.   A Matching function which evaluates the relevancy of the documents with respect to the user query.

The IR models can be categorized as Classical Information Retrieval models, Non-Classical Information Retrieval models and Alternative models for Information Retrieval.

2.3.1   CLASSICAL MODELS OF IR

This is the simplest model to build an Information Retrieval system. This model is based on the well-recognized and easy to understood knowledge of mathematics like probability. Classical models are easy to implement and are very efficient.

The three classical models of information retrieval are:

a. Boolean model;
b. Vector space model; and
c. Probabilistic models.

2.3.2   NON-CLASSICAL MODELS OF IR

Non-classical information retrieval models are based on principles like information logic model, situation theory model and interaction model. They are not based on concepts like similarity, probability, Boolean operations, etc. on which classical retrieval models are based on.

2.3.3   ALTERNATIVE MODELS OF IR

Alternative models are advanced classical IR models. These models make use of specific techniques from other fields like Cluster model, fuzzy model and latent semantic indexing (LSI) models.

2.4   BOOLEAN RETRIEVAL MODEL

This is the simplest retrieval model which retrieves the information on the basis of the query given in Boolean expression. Boolean queries are queries that uses And, OR and Not Boolean operations to join the query terms. For example:

Q1: Sachin AND Dravid.

The main shortcoming of this model is that it requires Boolean query instead of free text. The other disadvantage is that Boolean information retrieval model cannot rank the documents on the basis of relevance with the user query. It just gives the document if it contains the query word, regardless the term count in the document or the actual importance of that query word in the document.

Example:

Document collection:

d1 = "Sachin scores hundred."
d2 = "Dravid is the most technical batsman of the era."
d3 = "Sachin, Dravid duo is the best to watch."
d4 = "India wins courtesy to Dravid, Sachin partnership"

Lexicon and inverted index:

Sachin   →   {d1,d3,d4}
score   →   {d1}
hundred   →   {d1}
Dravid   →   {d2,d3,d4}
technical   →   {d2}
batsman   →   {d2}
watch   →   {d3}
India   →   {d4}
partnership   →   {d4}
win   →   {d4}

Result set:

{D1, D3, D4} ∩ {D2, D3, D4} = {D3, D4}

In Boolean model, the IR system retrieves the documents based on the occurrence of query key words in the document. It doesn't provide any ranking of documents based on the relevancy.

2.5   VECTOR SPACE MODEL

In vector space model documents and queries are represented as vectors of features representing terms. Features are assigned some numerical value that is usually some function of frequency of terms. Vector Space Model of Information Retrieval provides rankings of the resulted documents based on the similarity of the query vector with the documents vector. That is, it provides documents in the order of relevance with the user query.

In VSM, each document d is viewed as a vector of tf´idf values, one component for each term

So we have a vector space where

a. terms are axes; and
b. documents live in this space.

$$\vec{d}_j = \left(t_{1,j}, t_{2,j}, \ldots, t_{n,j} \right)$$

$$\vec{q}_k = \left(t_{1,k}, t_{2,k}, \ldots, t_{n,k} \right)$$

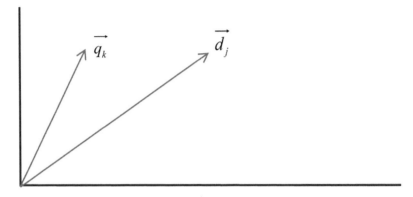

FIGURE 2.5   Representation of document and query features as vectors.

Ranking algorithm compute similarity between document and query vectors to yield a retrieval score to each document. The Postulate is: Documents related to the same information are close together in the vector space.

2.6   PROBABILISTIC RETRIEVAL MODEL

This model works in following phases:

In first phase, some set of documents is retrieved by using Vector Space Model or Boolean model.

In next step, the user reviews these documents produced in phase 1, to look for the relevant ones and gives his feedback.

Finally the Information Retrieval system then uses this feedback information to refine the searching criteria and rankings of the retrieved documents.

This process is repeated, until the user gets the desired information in response to his needs.

2.7   METRICS FOR IR EVALUATION

The aim of any Information Retrieval system is to search document in response to a user query relevant to his information need. The performance of IR systems is evaluated on the basis of how relevant documents it retrieves.

Relevance is subjective in nature. It is dependent upon a specific user's judgment. The true relevance of the retrieved document can be judged by the user only, on the basis of his information need. For same query statement, the desired information need may differ from user to user. Hence the same retrieved document may be useful for user and non-relevant for another one. Only the user can judge the relevance of the retrieved document according to his need. However measurement of this "true relevance" is not feasible.

Traditionally the evaluation of IR systems has been done on a set of queries and test document collections. For each test query a set of ranked relevant documents is created manually then the system result is cross-checked by it.

Different performance metrics are used to assess how efficiently an IR system retrieve the documents in response to a users information need.

The common parameters used for evaluation of an IR system are:

a.   Coverage of the collection
b.   Time lag
c.   Presentation format
d.   User effort
e.   Precision
f.   Recall

Effectiveness is the performance measure of any IR system which describes, how much the IR system satisfy a user's information need by retrieving relevant documents.

Aspects of effectiveness include:

a.   whether the retrieved documents are pertinent to the information need of the user.
b.   whether the retrieved documents are ranked according to the relevance with the user query.
c.   whether the IR system returns a reasonable number of relevant documents present in the corpus to the user, etc.

2.7.1   RECALL

Recall is one of the effectiveness measures of an IR system. It is the total number of closely related documents return by the IR system, when a user gives a query. Among all the relevant documents present in the corpus, the number of relevant documents system extracted from the corpus is called the recall.

Mathematically recall can be expressed as:

$$\text{Recall} = \frac{(\text{number of relevant documents retrieved})}{(\text{Total number of relevant documents present in the corpus})}$$

2.7.2   PRECISION

Precision is another common parameter, used to evaluate the IR system. Among all the retrieved documents, the number of documents, which satisfy the user's information need, is called the precision of the system.

Mathematically precision can be given as:

$$\text{Precision} = \frac{(\text{number of relevant documents retrieved})}{(\text{Total number of documents retrieved})}$$

Trade-Off Between Recall and Precision

If system returns the documents even if they have very low relevance to the query statement but contains some query words, then the Recall value of the system increases but the precision decreases because of lot of junk documents.

If the precision is very high, i.e., system return the documents having very high relevance with the query statement, then in this case there are possibilities that the IR system may skip some documents that may be useful for the user, i.e., Recall decreases (Figure 2.6).

2.7.3   F-MEASURE

Another parameter is F-measure, which consider both the precision as well as recall for measuring the effectiveness of the IR system. It calculates the harmonic mean of recall and precision.

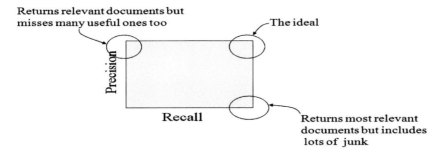

FIGURE 2.6   Trade-off between precision and recall.

Mathematically F-measure can be given as:

$$F = \frac{2PR}{P+R},$$

where P denotes the precision of the IR system and R is the recall value of the IR system.

2.7.4   E-MEASURE

E-measure is an enhanced version of F-measure. E-measure allows weighting emphasis on precision over recall.

Mathematically it is defined as:

$$E = \frac{\left(1+\beta^2\right)PR}{\beta^2 P + R} = \frac{(1+\beta^2)}{\dfrac{\beta^2}{R}+\dfrac{1}{P}}$$

The value of b depends on these criteria:

a.   To give equal weights to precision and recall, set b to 1. In this case (E = F).
b.   To give more weight to precision set b>1.
c.   To give more weight to recall set b<1.

2.8   PROMINENT IR TOOLS

There are various tools available in the market which users can explore and use for their applications.

2.8.1 NETOWL [6]

NetOwl is an advanced information retrieval system with automatic indexing and summarization capabilities. It is easy to use and also cost-effective for common users. NetOwl by identifying key concepts and relationships allows users to quickly get the required information according to the need by eliminating irrelevant contents. This suite provides products which can be used for variety of IR tasks like multilingual entity extraction, cross lingual Named identity, entity and aspect based sentiment analysis, text mining and identity resolution.

2.8.2 TREC_EVAL [7]

TREC_EVAL is software designed for evaluation of various information retrieval systems.

2.8.3 MALLET [8]

MALLET is a Java-based package. It can be used for various statistical NLP tasks like document classification, clustering, topic modeling, information extraction, and other machine learning applications to text.

2.8.4 LEMUR [9]

The Lemur Project develops search engines, browser toolbars, text analysis tools, and data resources. Some of the prominent tools developed by the Lemur project like Indri 5.15, Galago 3.16, RankLib 2.12 and Sifaka 1.6 are very useful in research and development of IR software.

2.8.5 APACHE LUCENE [10]

The Apache Lucene™ project develops open-source software like Lucene™ core (a powerful java library for indexing, spellchecking, analysis/tokenization and search.) and Solr™ search server.

2.8.6 TERRIER [11]

Terrier is open source highly flexible search engine. It provides an ideal platform for the rapid development and evaluation of large-scale retrieval applications

2.8.7 EUROSPIDER

The EUROSPIDER is commercial version of the IR system SPIDER, developed by the Swiss Federal Institute of Technology. This Information Retrieval (IR) system can be effectively used for very large and complex data collections to search for relevant information.

This advanced IR system provides various features such as relevance ranking, feedback searches, linguistic document analysis, and automatic indexing.

2.9 FUTURE DIRECTIONS

Researchers have done a lot of work in the field of information retrieval, but the main focus of the IR community was on the English and European languages. A huge volume of data related to almost all the things in the world is available on internet in different languages. Very little percentage of people is actually using English as their main language. The utilization of information available in other languages such as Asian language is not as efficient because the lack of tools and techniques. So there is a need of some efficient tools and techniques to represent, express, store and retrieve the information available in different languages.

A number of information retrieval systems are available for English and some other European languages. Work involving development of IR systems for other languages is only of recent interest. Development of such systems is constraint by the lack of the availability of linguistic resources and tools in these languages. The main obstacle the IR community faces in the development of efficient information retrieval systems for these languages is the lack of availability of resources, tools and techniques.

The study shows that most of the work done by the researchers is for text based IR systems. However now with the exponential use of internet services we have a huge amount of information in different media formats like audio, images and video. So there is emerging need of development of such IR systems which can effectively support different types of data.

2.10 CONCLUSION

We have huge amount of information available on internet consisting of billions of documents. For efficient access of this information there is a need of Information Retrieval systems which can effectively respond to the user's query and satisfy the information need of the user. This chapter introduces and defines the basic IR concepts. It also provides knowledge about the concepts like stemming, indexing and stop word removals which is very much essential to know while developing an IR system. The chapter discussed about different models and techniques which one can incorporate for building an Information Retrieval system. The main focus of this chapter was on classical IR models as they are widely used and easy to implement. Evaluation techniques for IR systems were also briefly discussed. Finally a research direction is provided to work upon in future.

KEYWORDS

- **Boolean model**
- **indexing**
- **IR models**
- **IR system**
- **stemming**
- **vector space model**

REFERENCES

1. Husain, M. S., & Siraj, I., (2013). A language independent approach to develop URDU IR system. *International Conference on Computer Science, Engineering and Application Technology (ICCSEA)* (pp. 397–406). Delhi: CS & IT-CSCP 2013.

2. Mooers, C. N. (1950). "The theory of digital handling of non-numerical information and its implications to machine economics," in *Association for Computing Machinery Conference.*

3. Shahid, H. M., (2012). An unsupervised approach to develop stemmer. *International Journal on Natural Language Computing (IJNLC)*, 15–23.

4. Husain, M. S., (2013). A language independent approach to develop URDU stemmer. In: *Advances in Computing and Information Technology* (pp. 45–53). Springer.

5. Shahid, H. M., (2013). An unsupervised approach to develop IR system: The case of URDU. *International Journal of Artificial Intelligence and Applications (IJAIA)*, 77–87.

6. "Featured Products," SRA International, Inc., 2020. [Online]. Available: https://www.netowl.com/ (accessed on 25 February 2020).

7. "trec_eval," NIST, [Online]. Available: https://trec.nist.gov/trec_eval/ (accessed on 25 February 2020).

8. "Mallet," [Online]. Available: https://github.com/mimno/Mallet (accessed on 25 February 2020).

9. "The Lemur Project," [Online]. Available: lemurproject.org (accessed on 25 February 2020).

10. "Welcome to Apache Lucene," Apache Software Foundation, [Online]. Available: https://lucene.apache.org/ (accessed on 25 February 2020).

11. "Welcome to the Terrier IR Platform," University of Glasgow, [Online]. Available: http://terrier.org/ (accessed on 25 February 2020).

12. Hersh, D. W. M. F. M. E. W. R., (2001). "Information Retrieval Systems," in *Medical Informatics. Health Informatics*, New York, NY, Springer.

13. William B. Frakes, R. B.-Y. (1992). Information Retrieval: Data Structure and Algorithms, Prentice Hall.

14. Berthier Ribeiro-Neto, R. B.-Y. (2011). Modern Information Retrieval, Addison Wesley.

15. Bates, M. J. (2011). Understanding Information Retrieval Systems: Management, Types and Standards, CRC Press.

16. Korfhage, R. R. (2006). Information Storage and Retrieval, Wiley India Pvt. Limited.

17. Tanveer Siddiqui, U. S. T. (2008). Natural Language Processing and Information Retrieval, OUP India.

18. Shahid, H. M. (2012). Language Independent Approach to Develop Information Retrieval System: A Case Study of Urdu Language.

CHAPTER 3

Futurity of Translation Algorithms for Neural Machine Translation (NMT) and Its Vision

K. MANDAL,[1] G. S. PRADEEP GHANTASALA,[2] FIROZ KHAN,[3]
R. SATHIYARAJ,[1] and B. BALAMURUGAN[4]

[1]*Assistant Professor, School of Computing Science and Engineering,
Galgotias University, Uttar Pradesh, India,
E-mails: kuppanmandal@gmail.com (K. Mandal),
sathiya.peace@gmail.com (R. Sathiyaraj)*

[2]*Professor, Department of Computer Science and Engineering,
Malla Reddy Institute of Technology and Science, Hyderabad,
Telangana, India, E-mail: ggspradeep@gmail.com*

[3]*Professor, IT Faculty, Dubai Men's College, Higher Colleges of
Technology, UAE, E-mail: fk7@hotmail.com*

[4]*Professor, School of Computing Science and Engineering,
Galgotias University, Uttar Pradesh, India, E-mail: kadavulai@gmail.com*

ABSTRACT

Machine Translation is the automated method of conversion text from one language to other foreign languages by using intelligence approach of automated algorithms, when machine learning concepts applied to it. Over the conversion process, the mechanism utilized with intermediate process to translate from one source language to a foreign language. It is the part of artificial intelligence (AI), when intelligence has the prediction capability makes as machine learning and much accuracy, known as deep learning. Conversion of one form to another form is called translation that makes trending as a world as a single language, that form of automated conversions called machine translation. This kind of translation requires the background knowledge about the both languages or for multiple languages, in this way a classical machine translation needs to satisfies a set of constraints based as guidelines. Because of some languages produces same text over different meanings over part of utilization in the context and, sometimes provides same meaning over different words. So, the basic two methods of classification for machine translations were statistical and neural machine translations (NMTs). In the traditional method, usage of statistical machine translation (SMT) approach to perform translation was the way to predict possible best outcome with definite algorithms. But in NMT, approach applies the dynamic algorithms for best predictability of word on translation according to the context appropriately. In the chapter, the proposed assertion explains about all form of NMT with complete translation of process involves neural network, that produces more number of accessible input to find the best possible output according to utilization. This kind of translation applies multiple strategies on different stages for translation, Stage one implements the translated based on word with complete sentence and, Stage 2 implements the translated on model over the word within the sentence context. The solidity of NMT allows the learning ability over point-to-point bases on the background knowledge input to the predicable target output. And also brief discussion on various kinds of NMT conversion principles such as like Google neural machine translation (GNMT), open source neural machine translation (OpenNMT), deep neural machine translation (DeepNMT), and so on.

3.1 INTRODUCTION

Since the human existence on the earth, allow to create a medium to share information through any mode among them. By that time, language has been evolved for the common to communicate with each other. Even all the living organisms such as plants and animals also being communicate with each other, within their communication medium [1]. And also, scientist has proved that plants also communicating with each other as well as helping each other. Similarly, an animal makes sounds to communicate but not all animals produces same sound. When human starts civilization sound becomes language and, they were become groups living in individual locality requires to communicate within the group also with others [2].

After so many evolutions, human were settled at different parts of the world. That evolution makes changes over the language for the communication, at last if they want to communicate with each other there was a problem with different languages among the groups. So, people need to be known of the both language to communicate amicable between the groups. The method of converting a language from one to another foreign language with similar meanings to arrive, the process is called translation. Here people undergoes learning of two languages and its importance, consumes much amount of time to become familiar and fluency among the languages for translation [3].

Later than, the invasion of technology in translation brings machine into the process of converting one language to foreign language in an automated way is known as machine translation [4]. Machine translation is not a technology rather, than it is a methodology. Machine Translation is the automated method of conversion text from one language to other foreign language by using intelligence approach of automated algorithms, when machine learning concepts applied to it [5].

3.1.1 APPLICATIONS OF MACHINE TRANSLATION

1. **Text-to-Text Translation:** It was the basic form of machine translation, for this applying automated mechanism on conversion of text from source language to target language requires the meaning of phrase and analysis synonym apply to the target language suitable and similar meaning translation [6]. For example, few words in source language has same spelling and pronunciation but according to the situation it might be make sense of the word like in English *"bat* fly" and *"bat* was broken" from the two phrases bat has same spelling and pronunciation but as of usage it makes animal and thing accordingly [7].

2. **Text-to-Speech Translation:** It converts the data into a sound like somewhat related to play music in a machine. This means the raw data of text will be loaded to a database of both letters and words in a particular language and then makes to sound for each, in order to achieve pronunciation well good enough [8]. For an example of converting the text of one language to the same language speech, then it requires less about of database to make type conversion from digital data to analog data. Also in English few words have alphabets with silence in pronunciation like *"pseudo"* as \'su-(,) do\ here, 'p' alphabet was silence at pronunciation. Even making, syllabifying for pronunciation also difficult to map alphabets in a

word. But in the case of different language translations, it was a little bit more complex of translate to target language along with pronunciation [9].

3. **Speech-to-Speech Translation:** It was the method to translate of speech source language sound to speech of target language sound [10]. In this conversion, the pronunciations of words may differ basically among languages which made the conversation difficult, that too in speech-to-speech translation source language sound signal need to analysis and conversion of target language by synthesis input acquired from analysis output with huge volume of data transaction with knowledge base predict the appropriate word has correct meaning [11]. For example, the translation of the English language of the word having *"bear"* and *"beer"* as \'ber\ within a sentence makes it much different while converting to any other language. But pronunciation might not differ as much as so, this translation requires the complex data in neural networks (NN) to obtain all the possible translation with similar data and phrases to extract correct meaning [12].

4. **Speech-to-Text Translation:** It has the complex task of translate, because the pronunciation of words from the different origin peoples has distinct ways. Then the multiple nodes of networked words need to choose the appropriate word as of the sentences, phrases, and grammatically. For example in the English language the word *"hear," "ear," "year"* and *"here"* as \'hir\ most probably has same sound of human vocal but as of the situation it differs and may not be possible either, or, these words [13]. In this case, when the translation of one language to another language was an embarrassing task; like analyzing the analog signal of English speech synthesis to native language English text with the help of knowledge base, then apply the translation mechanism to another foreign language conversion [14].

Some more applications in real-time uses of machine translation as follow:

- Machine translation was enhanced to checking of grammatical errors in the target language also in the platform for writing new scripts.
- It provides the better environment for human-computer interaction and vice versa through natural language processing (NLP) for visually impaired humans.

- Machine Translation reduced the time consuming to conversion of language as in a simplest way of pop-ups for translation.
- It makes flexible to understand human language and computer language allows converting huge content of data such as e-books and websites in a structured data and stored in conventional database.

3.2 OVERVIEW OF MACHINE TRANSLATION

Nowadays more number of organizations are involved in machine translation; because of huge content of data conversion requires accuracy and speed of conversion from the source language to the target language. Using various methods of analyzing the pros and cons in machine translation to tackle the problems of language conversion [15]. While translating the grammar part is more important, so that it may change the entire meaning of the phrase. Such as, in some languages, verbs utilize only at the end of the sentence but some other language has in between the sentence [16]. Machine Translation is the part of artificial intelligence (AI), when intelligence has the prediction capability makes as machine learning and much accuracy, known as deep learning. Conversion of one form to other form is called translation that makes trending as a world as single language, that form of automated conversions called machine translation [17].

In the machine translation system, the specialized methodology based on time constraint, economically feasible and end-user [18]. Also, favorable technology provides the machine translation more useful such as like translation memory (TM) and optical character recognition (OCR). TM is the collection of linear data that was stored as "chunks," those chunks like phrases, sentences, paragraphs or a unit of words which was translated from one to other by human translators. Translation unit has translation of source language in storage of TM [19]. OCR is one of the most efficient technology to convert the images to text format in an automated way. Those scanned content in the images will be clear an error-free on translation, but even with OCR technology also the major disadvantage. When the image does satisfies the color contrast recognition constitutes difficult to convert the text [20]. Furthermore, the number of organizations involved in machine translation using the assets of TM will undergo the process of conversion and also as well as many learning segments from the human linguistic content. The overall process provides the quality of deliverables of translation benefit significantly [21].

3.2.1   MACHINE TRANSLATION SYSTEMS

Machine Translation is a significant technology in the growing aspects of digital world. It bridges the gap between the common communication medium for different languages as in a unit. But in the traditional methods requires huge effort to translate also being a cost-effective and vast amount of time-consuming [22]. At the initial stages of there were significant challenges encountered. After undergoing so many iterative experiments and recursive modifications with the latest technology of NLP such as it includes more number of studies of language computation, lexical analysis, linguistic manipulation, computer engineering, AI and others [23].

3.2.1.1   GENERIC MACHINE TRANSLATION

When the machine translation is not allowed to set up a user-defined specific stream, and that kind of translation termed of generic machine translation. Most of the open-source translation organizations for common to all were following the generic method. This will allow to translate often produce on a piece of text but not in a grammatically and not follows other streams. Without human intervention for publishing context of translated content was impossible in generic machine translation [24]. Even many organization has an issue in this translation method encountered bad name for translation also sometimes it requires more work for editing on translated content.

   Giant technological firms were offering services through generic machine translation, but misassumption on hopes of translation quality. This leads the research and development on translation in a quality of translation in a customizable way of machine translation [25].

3.2.1.2   CUSTOMIZABLE MACHINE TRANSLATION

Customizable Machine Translation is provided by one of the giant organization in AutoML by Google. In which the cloud platform allows to do their own translation data sets to customize trained domains on machine translation different models with current tech of NN [26]. As in AutoML also with more specific domain on AutoML Natural Language and AutoML Translation, offering more custom on domain-specific model on NMT. The customizable process allows new data sets to even provide more appropriate translation as in economical costing on translation APT to NMT engines supports in bidirectional [27].

There was more number of challenges to attend translation API output with the NMT engine. It potentially allows many organizations to build custom engines for translation mostly relies on language data brokers. In distinct machine translation model has many forms to process, using technology to compare the competitors by offering more generalist customers instead of programmers, to point out the neural machine translation (NMT) output quality always depends on the hardware, comparatively with software process. This provides an advantage of the Google cloud platform provides the best solution in customizable machine translation method [28] (Figure 3.1).

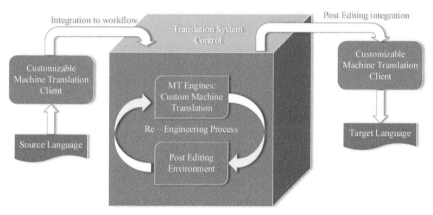

FIGURE 3.1   Customizable machine translation.

3.2.1.3   ADAPTIVE MACHINE TRANSLATION

Adaptive machine translation is really a new technology that allows the machine translation system to get corrections from the input during the process of regularization. In statistical machine translation (SMT), the system provides concept as an adaptive method in SMT has two major components of translation model and language model in translation model [29]. It is likely by linguistic text of a phrasal method that provides weight or score of the context, which has been probably piece for translation having high value. This was attained by not a single method used many translations was generated to occur the target language with the help of language model [30]. The language model is a significant for collection of all the text in mono linguistic of targeted language, which increases chance of good translation. Now apply the traditional method of teach the system in a better way by training it. This produces a good quality of translation based on the phrase, which achieves its course of weight allow the system to be built up for long time and also vast amount of data with high processing speed for translation. But training a system will takes a long time period in the system, identically have to machine translation will have a post-editor correction method implementation to translate with high quality by the following sequence of steps:

- **Step 1:** Machine translation generated output.
- **Step 2:** Correction received from the user.
- **Step 3:** Launch the correction with the machine translation system.
- **Step 4:** It reduces the same error will occur again on the output.

It works with static and dynamic methods, applicable on larger data set collection to generate translation and updating of data with post-editing of learned language. Adapting the system makes new corrections on the data sets with the weight accessing method of algorithmic perspective on language models [31] (Figure 3.2).

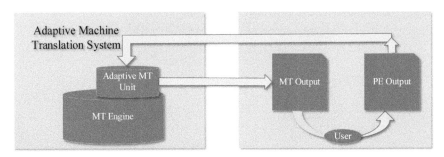

FIGURE 3.2   Adaptive machine translation.

3.2.2   *CATEGORIES OF MACHINE TRANSLATION*

Among different categories, the most widely used machine translation on rule-based which was used for commercial purpose for translation follows different strategies dictionary has the feature of obtaining the matching word for target language, transfer based successful transfer of source to target translation with reference datasets, and interlingua also different from existing defines the translation between multiple target language in a unit time of conversation [32]. The popular two methods of classification for machine translations were statistical and NMTs. In the traditional method, usage of the SMT approach to perform translation was the way to predict possible best outcome with definite algorithms. But in NMT, approach applies the dynamic algorithms for best predictability of word on translation according to the context appropriately [33].

3.2.2.1 RULE-BASED MACHINE TRANSLATION

Rule-Based Machine Translation is simply an acronym as RBMT; it is one of the Classical Approach of Machine Translation and 1st machine translation system for commercial purposes. RMBT approach is based on semantic (language-producing) information basically it can process information from source to target languages whether it maybe unilingual, bilingual or multilingual and it asylums different morphological, syntactic, and semantic symmetries of all languages [34]. RBMT generates target language in the same or different form based upon source language (Figure 3.3).

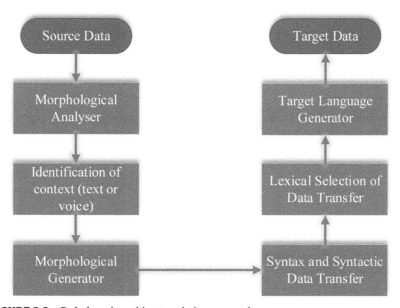

FIGURE 3.3   Rule-based machine translation approach.

Stages of RBMT include:

- Morphological analysis of source code;
- Parsing source language into syntactic groups;
- Getting syntactic information about each group;
- Dictionary-based translation.

3.2.2.1.1 Dictionary-Based (Direct) MT

In dictionary-based machine translation (DBMT), the translation is possible by word-by-word, here the dictionary lookups may be done with or without morphological analysis. Sometimes manual translation will be done more quickly for correcting the grammar and syntax. The DBMT systems are normally used for cross-language retrieval systems (CLRS). Because these are the systems capable of producing the required target language from the given source language [35].

For example, Source language is English and the target language is Telugu, Tamil, and Hindi, etc. Some researches have been proved that cross-lingual information retrieval (IR) system for the agricultural society for farmers in Tamil Nadu which translates Tamil to English which uses morphological analyzer for accessing information related to farmers in their native language Tamil via the available language English [36] (Figure 3.4).

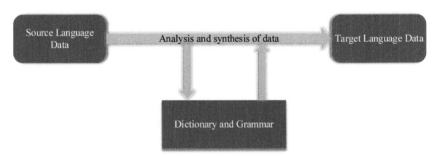

FIGURE 3.4   Dictionary-based machine translation.

3.2.2.1.2 Transfer-Based Machine Translation

The transfer-based machine translation is used to generate the source language into their target language in different steps. First, it analyzes the given input data for morphology and syntax and also sometimes, it searches for semantics. Second, it creates a structure for the input data for transferring the related output. Third, with the help of input data, the corresponding structure for output is created and finally, the output target data is generated as per the specified dictionaries and grammatical rules [37] (Figure 3.5).

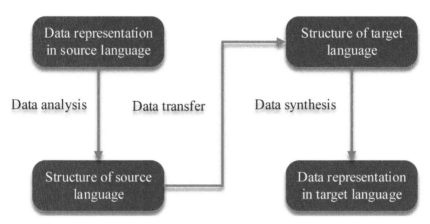

FIGURE 3.5   Transfer-based machine translation.

3.2.2.1.3   *Interlingual Machine Translation*

Interlingual is an artificial language that can be used to indicate in any usual language. It acts as an intermediary of more than two languages. It takes text as the main source for generating one language into their equivalent language [38]. And also, it shows the structural difference between languages because a single word having more than one significance in other languages (Figure 3.6).

3.2.2.2   *EXAMPLE-BASED MACHINE TRANSLATION*

Example-based machine translation acronym as EBMT which can be used for translation one language into another language with the help of steps matching, alignment, and recombination [39].

- **Matching:** It finds the target language.
- **Alignment:** In this step, it extracts corresponding match.
- **Recombination:** Here it produces the final translation.

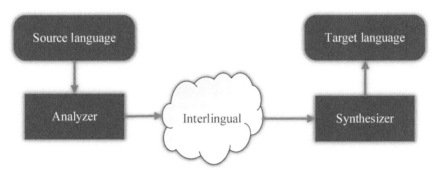

FIGURE 3.6   Interlingua machine translation.

For example, let us consider the source language is English the sentence "my name is Pradeep" their corresponding target language in Telugu is "నా పేరు పరదీప్" [40].

The structure of example-based machine translation is as follows (Figure 3.7):

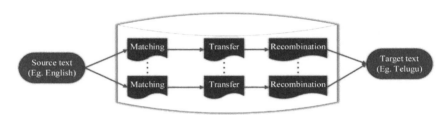

FIGURE 3.7   Example-based machine translation.

3.2.2.3   *HYBRID MACHINE TRANSLATION (HMT)*

Hybrid machine translation (HMT) is a type of machine translation; it uses both the combination of rule-based machine translation and SMT. i.e., it extracts both the features of RBMT and SMT [41]. The main characteristic of HMT is to combine multiple machine translation approaches into a single machine translation system; by doing like this if any failure occurred in any single technique can also leads to accomplish an acceptable level of precision.

By using HMT English to Arabic, machine translation has been developed with the combination of RBMT and EBMT [42] (Figure 3.8).

FIGURE 3.8 Hybrid machine translation.

3.2.2.4 STATISTICAL MACHINE TRANSLATION (SMT)

SMT is also a kind of machine translation by using statistical models SMT systems can be obtained [43]. SMT methodology makes use of different statistical translation models produced by the analysis of unilingual and bilingual training data. By using most commonly occurrence words or sentences or expressions or phrases the SMT can be takes place by using SMT algorithms. E.g., decoding algorithm [44] (Figure 3.9).

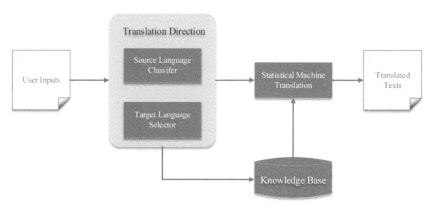

FIGURE 3.9 Statistical machine translation.

The most general challenges related to SMT are:

1. **Arrangement of Sentence:**
   - Arrangement of words.

2. **Statistical Deviation of Data (or) Statistical Occurrence of Data:**
   - Words or sentences which are occurred in different form that is unable to identify properly.
   - Speech or voice reorganization is mandatory because the pronunciation of words or usage of a particular word can vary depends on the usage.

3.3   NEURAL MACHINE TRANSLATION (NMT)

NMT is quite opposite to SMT, because it has foundation of neural network technology. NMT is the powerful approach of Machine Learning and AI. NMT is the trending technology in machine translation, where the system developed by NN to produce eloquent translation as compare to existing method, also increasing the level of accuracy and quality of machine translation [45].

NMT is the method of applying the mechanism of neural network for translation in an automated process. In neural translation, the input as like data format in text of parallel was aligned in source file line-by-line for target translation. Lines in the source file contain tokens of words and that spited into segments of sentence as tokenize for translation. In learning the data for translation also having defects, as of NMT the output was framed using probability distribution on words for prediction [46]. If the input contains misspelled of words may not be found in a dictionary of its translation vocabulary, then the possible words in the vocabulary has infinite, using the artificial method of limiting the size of possible words to handle through our model. Also, another drawback of data training with the model has some root word as derived with the other forms in words contains additional of some letters at the end, but the stem of the words was common [47].

NMT is one of the novel variants in corpus-based machine translation (CBMT). In general, corpus means a collection of data it may be sentences, words, text or it will be in any format, because of it CBMT is also to be referred as data-driven machine translation (DDMT) or corpus-driven machine translation (CDMT). A CBMT is similar to SMT it also translates enormous amount of data at a time which contains different segmentation of sentences from source to destination [48].

3.3.1   HISTORY OF NEURAL MACHINE TRANSLATION (NMT)

NMT has the term because of the Neural Network mechanism included in Machine Translation. During 2014 the mechanism of neural network appended in machine translation, later on, more researches were carried to greater end [49].

NN in NMT is a different methodology in which a distinct, large neural network is competent, exploiting the translation performance. Neural network belongs to a part of deep learning technology which indicates the interconnection of neuron cells that permits to distinguish patterns or shapes, it can learn, it can solve problems and allows taking decisions. It uses artificial brain to take these kinds of actions that's why it is also called as artificial NN or artificial brain NN. A NMT takes input as thousands of artificial entities stimulate and generate their corresponding output entities [50].

3.3.2 NEURAL NETWORKS (NN) OF MACHINE TRANSLATION

Unlike in SMT has expression-based translation system which consists of small parts of data in huge chunks in individual, NN system initiate to build learn and train a single model, similarly large NN employ to read passage and produced the corrected translation as output [51].

NMT has the inspiration of human brain functionality incurred of processing units as neurons like neural units which are connected to each other in NN. As a natural behavior of neurons connected of two states in basic were input and output, in illusion of chance the states has inputs was connected to N neurons produces mixed up networks. In detail, the various states of neurons were connected at each stage of nodes get multiplied by weights of the corresponding neurons which was connected. Later stage the output was obtained through the activation function on input neurons with weight [52] (Figure 3.10).

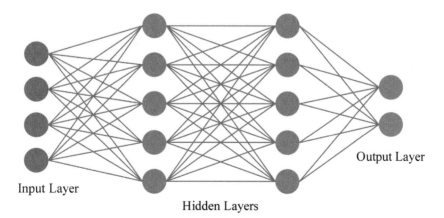

FIGURE 3.10 Neural networks.

Similar of neuron process in the NN was grouping units of distributed representation at various layers. As like in NMT, words, phrase, sentence were processed in a parallel of nodes in distributed method allows to form a large set of trained neurons to develop distributed representation of the content for better-trained words, phrase, sentences to derive the actual out through the activation function on different layers [53]. As illusion that, the words and sentence has been nodes of NN, where two identical words and sentences are close to nearby with identical coordinates, similarly the distinct words and sentence were far apart to each other along its coordinates. The defined process in two dimensions was not enough to predict efficiently. Considering of multi-dimensional perspective accommodates the relationship between the identical and distinct of words and sentences in machine translation. In NN the nodes involved in the connection was more, like the node in a layer was linked to all the nodes in previous as well as preceding layer nodes. Henceforth, defines the processing method involved in much shallower and termed as deep learning [54].

3.3.3   ENCODER-DECODER MINIATURE

In encoder-decoder, the process entails recurrent NN, which is running in two directions to determine as a bidirectional recurrent neural network. In the encoder part, the method which has been involved to the input language acquired from source has been embedding into a matrix with a sequence of words and, then it has been processed through recurrent neural network. This process produces a result for the hidden layer of a neural network with left context and the right context, from the beginning and end the sentence [55]. As well as in the decoder part, the context of encoder's hidden layer recurrent neural network in the form of left and right context will generate a new hidden layer of decoder to create output word prediction in the form of probability distribution with overall possible vocabulary. If we consider our vocabulary has 5000 words, then the production of output with probability distribution gives 5000-dimensional vector for each word in the vocabulary [56] (Figure 3.11).

Encoder-Decoder miniature process for translation has a powerful design and to manipulate the different length of information from input to output. The source language used the encoder path to create fixed length of chunk sentence representation, which allows into decoder path building an output translation from encoded chunk of sentence. The main aim of this system, to escalate translation of languages and increase its probability of accuracy has given through translation system [57].

welcome → Encoder → RNN → Decoder → q ı{ q † «

FIGURE 3.11   Encoder-decoder architecture.

3.4   WORKING PROCESS IN NEURAL MACHINE TRANSLATION (NMT)

The working method of NMT speculates different phases; *Training* of neurons can be taken in a fast-growing environment, because the source data in one form the corresponding obtained target data in another form. In neural translation system, the data will be translated into different sets of groups it follows distributed representation concept sometimes it is also called as a gold standard translation for the reason that it follows hard-bound rules and semantics for converting source sentence to target sentence example English to French (EN-> FR). By training the neurons (artificial brain), one can easily identify the connections between them and also the obtained output [58]. By training of corpora in NMT, we can easily process thousands and millions of sentences, words, and so on. This kind of NMT is better compared to old SMT.

By using training algorithms such as Gradient descent, Newton's Method, Conjugate gradient, Quasi-Newton method, Levenberg-Marquardt algorithm we can identify the error function or loss function because at the time of training there is a possibility of changing the weights of a neuron at the time of translation. By using these training algorithms one can able to avoid much changes in the values so that it can be helpful to avoid the loss or errors taking place while translation. For checking the possibilities of data at the time of conversion, the training of data is mandatory. The major process of machine translation was to identify the *next word predictive* methodology. NMT uses encoding and decoding schema for translation purposes [59]. NMT utilizes text completion device for indicating the source sentence, at the time of source generation it uses an encoding system for text completion device. At the time of conversion, the corresponding target terminology of sentence is identified. With the help of decoding system, the corresponding targeted sentence can be identified in their native language form. Sometimes one words having different forms or meaning in another language that is the condition has to interact clearly while translation. For this purpose, it uses a particular type of interactive machine translation or interactive translation prediction. These translation completion predictions are suggestible in SMT for accepting the target text [60].

The vital process of translation always distinct from words to sentence form of representations; similarly, NMT follows neural computation because for translating a single sentence, it will divide into individual modules of words or sub-words. The representation of words must have follow carefully in order to generate the corresponding words or sentences. At the time of translation representation of words must be in properly computed, after computing each word compounds them and form a proper sentence. In a monolingual or multilingual environment, it is important to produce word or group of words from left to right or right to left as it was specified. This kind of representation of individual sometimes referred to as embeddings or continuous bag-of-words embeddings in distributed representation. Every language follows their own semantic and syntactic properties for representing word or group of words. For representation of words it follows a vector representation of neurons e('word') example e('hi').

Translation of neurons in NMT can be takes place with the help of encoding and decoding mechanism [61].

1. **Encode:** It takes place at source sentences or words.

    Let us consider a simple sentence for translating English to the Telugu language. It will follow the encoding schema at the source translation side [62]. Example: Source sentence in English form is 'My name is Pradeep.' This sentence has to convert into their corresponding Telugu language. At the time of encoding it follows a vector representation of individual words i.e., e('my'), e('name'), e('is'), e('Pradeep') and e('.'). The encoding vector translation will starts with the first word E('my') and then the encoder network combines E('my name') and later its successive E('my name is') and then E('my name is Pradeep'). Finally, it produces E('my name is Pradeep.'). This encoding process will take place either from left to right or vice versa.
    Whereas

2.   **Decoding:** It takes place at target sentences or words.
     Let's follow the decoding process for translation of the given
     source sentence, the given encoding sentence is E('my name is
     Pradeep.'). The decoding the sentence decoder contains two vector
     entities one is for holding the given encoded sentence D('my name
     is Pradeep," '), whereas second entity empty ' ' represents the
     corresponding target sequence sentence of words [63]. Consider
     x is the first position to hold target sequence p(x|'my name is
     Pradeep'). And the decoder to the Telugu word is x = 'నా' is the
     correspondig output. The decoder can be written as D('my name
     is Pradeep,' 'నా') and the possibity of vecotor probability can be
     shown as p(x|'my name is Pradeep,' 'నా') and the next position can
     be written as x = 'పేరు' is the output [64]. The sucessive decoder
     is D('my name is Pradeep,' 'నా పేరు') and the vector probability is
     p(x|'my name is Pradeep," నా పేరు') and then the next combination
     is D('my name is Pradeep,' 'నా పేరు ప్రదీప్') and the vector
     probability is p(x|'my name is Pradeep," నా పేరు ప్రదీప్'). The
     final decoded translated output for the target Telugu language is
     'నా పేరు ప్రదీప్.'

3.4.1   DEEP LEARNING

Deep Learning is the subset of machine learning and machine learning is
the subset of AI. Deep learning was inspired by human brain function of
neurons in the form of artificial NN [65]. The best part of NN has summing
up of vast amount data with even bigger models of pattern formulation and
more computational algorithm for distinct models will improves the techniques. Even backpropagation follows the multilayer perception networks
inartificial NN were described as deep learning in prediction. Deep learning
provides high computational power to the machine which can able to
recognize objects as well as language processing in real-time [66]. Finally,
it's getting too smart as AI deep learning provides trending technology
behind many real-time examples of driverless cars to distinguish between
different signboard on a road. And also, today's control of devices with the
help of speech recognize and image perception through electronic gadgets
are the good reason for performing well in deep learning, which are not
possible even before [67].

3.4.2   REPRESENTATION LEARNING

In supervised learning, the method to attend final output by providing a set of data contains predefined input and output able to provide a pattern between the input and output has a feature identifying unlabeled data by providing sufficient training to neural network, in this way representation learning emerged to greater extend [68]. In representation learning, apply the features of artificial neural network to represent the data for prediction to perform all the tasks done by machine learning has been achieved in a greater method through hidden layers of neural network. In representation learning is quite interesting, because it provides better prediction on supervised learning as well as other unsupervised and semi-supervised learning, it requires vast amount of data to train the data set of unlabeled data and also require a little amount of labeled data to train for better protection, this kind of representation learning provide more advantages comparative to human performance [69].

3.4.3   RECURRENT NEURAL NETWORK

Recurrent Neural Network is the methodology for prediction in AI to develop speech recognition and NLP to obtain the pattern with existing data sets for predictive analysis. This is one of the types utilized in deep learning with complex connectivity of brain neurons as like for knowledge representation of technology related to artificial NN [70]. To obtain a pattern between the input and output layers consists of number of intermediate layers of connected nodes avail the better pattern drafting method for predictions. Unlike traditional neural network of feed-forward mechanism but in recursive, neural network allows in both direction of backpropagation mechanism for pattern prediction of knowledge. Recurrent NN were used for sentence modeling at both directions. A bidirectional recurrent neural network has encoder at source language by neural network and decoder to predict the words using neural network at target language [71].

3.5   EVALUATION OF TRANSLATION

While comparing to the predecessor of machine translation should posses the better result on reliability represents not by single parameter, even it follows the measurement of both direct and automatic evaluation procedure. Even a few reputed firms were involved in translation and adopted NMT and being successful in their state [72].

3.5.1   *TRANSLATED BASED*

3.5.1.1   *EVALUATION IN AUTOMATED*

The measurement of evaluation in automated done by comparing the outputs of traditional model called reference translation with obtained output of machine translation on identical data matching with segments of sentence, often it provides advantages on NMT. Because of NN, inclusion methodology allows reordering of more sentences which has huge reference model than SMT [73].

3.5.1.2   *EVALUATION IN SUBJECTIVE*

Manual evaluation is otherwise called evaluation in subjective; it allows the measurement on the fluency of output. Consistent output obtained by NMT, which has more fluent comparative to SMT. Fluency in the sense of translation was different of exact prediction of translation language, it may not have similar meaning of source and target language [74].

3.5.2   *TRANSLATED ON WORD*

The post-editing method provides better translation accuracy but considers the post-editing time requirement. Post-editing effort was to perform the minimum amount of substation, unwanted word deletions, or necessary translated word insertions made in few words or in the entire sentence of machine translation gains the output in the exact meaning of target language with source language [75]. Still, the translation of NMT working in a finer way compare to the traditional process, but sometimes the preliminary results might not be satisfied as come to SMT. Recent trends of comparing the post-editing SMT with NMT predicted output contains the same data for both approaches, the measurement was based on editing time and post-editing effort for technical aspects based on a number of keystrokes utilized on the editing process. Not only with that, even has it quantified the number of keystrokes nonutilized words on translation. This projects the advantages of NMT over others on post-editing efforts based on keystrokes [76].

NMT will satisfy certain constraints during translation. Among the translation categories, NMT is recent and trending, which is being used by organizations and commercial usage of both online and offline through web service. This leads to avail the platform for research carried over machine translation [77].

1. **More Computational Power:** NMT requires more computation resources to train the data sets to compare to the SMT system. It requires parallel data sets for translation but not available for other organizations as well as small translation firms. To run these data sets requires high computing machines, so ordinary desktop computers couldn't computing efficiently or sometimes it takes time-consuming for translation applications [78]. There are number of machine translation engines which are most probably follow the NN mechanism for prediction. These engines required specialization skills to manipulate along with super powerful computing machines.

2. **Difference in Output:** The output obtained from NMT is always differs from SMT, even some time differs with other translation mechanism [79]. Because of huge data sets for translations on prediction apply to NN leads to mistranslation with actual language meaning, misplacing of translate words, etc.

3.6 CHALLENGES OF NEURAL MACHINE TRANSLATION (NMT)

NMT was the trending mechanism for translation of one language to foreign language in the dynamic world. Even latest mechanism also contains some flaws with it.

Here, some comparison to traditional statistical machine learning with NMT hold up the technology based on experimental results, in despite of that NMT has number of challenges need to overcome under less resources and out of knowledge [80]. So many issues with neural translation has to overcome with efficient translation such as like training data of the input domain language may not be satisfied with test sentence on training data in that case out of data from domain relies on amount of data for trained. For this problem prediction of entire sentence will results in unsatisfied prediction, rather than to predict the next word of the current word optimize the solution of that problem [81]. Although it contains other challenges when compared with the traditional ones, it provides more input suggestions for prediction best possible target language for input language through training data neural network. Even many iterations of translation data sets play a major role for conversion. The challenges of NMT were more in numbers of experiential reversion with SMT, which are listed below in six different aspects [82].

3.6.1 DOMAIN MISMATCH

A common challenge in translation is distinct for domain to domain, because words in a domain have different in meaning of its style according to the situation and usage. The challenge called out-of-domain, which makes bad on translation of religious and financial related statements while compare with traditional methods.

Some words have different translations and different meanings as of source language to target language translation in various styles done by the traditional translation system. The method of adopting NMT for domains used to train the data sets of domain system should lies in-domain data models, large amount of training data contains out-of-domain, but require to examine in robust performance on translation. Nowadays very few organizations were utilizing trained data sets by both SMT and NMT systems for all the domains to get better results of prediction for target language translation. For example as like captions under the video will be generated and translated directly by machine translation system, which provides more accuracy of training data set should be in-domain for prediction [83].

3.6.2 AMOUNT OF TRAINING DATA

NMT is the intelligence approach in translation which involves machine learning as common, this issue occurs often when low amount of dataset provided. But large dataset makes accuracy in prediction because of more data provides more linear form for translation. Some languages may not satisfy translation on less training datasets [84].

A common factor of traditional translation system requires large amount of training data produces the better result similarly. In SMT, system also produced better results when the training data gets doubled similarly, as like NMT also follows same rule to get better output through generalization. When, it has large content of their datasets of translation. For example, if the dictionary contains more words then follows of more synonyms provide similar meaning of the source word as well as along with the antonyms of the word provides better prediction for translation.

3.6.3   *RARE WORDS*

As like challenge with amount of training datasets, a similar issue with rare words. Even with high-influence language, being with huge dataset and the domain of rare words has less to translation accuracy [85]. This means rare words have less datasets of it, so there is a chance to mismatch of actual translation. In machine translation, pivot role of datasets to match appropriate words and its meaning to translate from source language to target language [86]. When the dataset increases the predictability also increases with exact translation, but less datasets of a particular word does not match similar meaning according to sentence sense [87]. For example, if a word did not occur frequently in the translation then using the same word for translation may be or may not provide appropriate translation to target language [88].

3.6.4   *LONG SENTENCES*

In few domains has lengthy sentence such as like in legal and others, also with complexity in a sentence makes toughness on translation. But as of now with NMT, the problem is encountered compare with traditional SMT [89]. Not only with often words, even with lengthy sentences also cause challenge to encounter, some text like very lengthy of religious and legal contextual may leads a single mistranslate with entire collapse of content meaning. And also with the major concern on poetic writing of literature may not have proper explanation, which leads to mistranslate content. For example, proverb has the set of words with definite form, but based on the situation its usage and meaning differ from place to place. Processing of such phrases has definition of paragraph of different sense of sentence [90].

3.6.5 WORD ALIGNMENT

The concentration of NMT was the assessments of output language alignment to inputs language done by probability distribution on the input define weight on the words of language of its representation. The role of word alignment was between the input language with output language, but not in the way of traditional machine translation. In word alignment, it uses the absorption of probability distribution on the words and sentence of language [91].

For consideration when the translation contains verb then it require more observation on it, because of subject and object main decides the action of that verb may disambiguate. Moreover, translation complicates the process of words applied on bidirectional of recurrent NN that provides representation of each word by the context of entire sentence. Now the alignment mechanism on the source and target language is crisp. Consideration, the proceeding works shows alignment done by the attention model of probabilistic distribution translations; in another method, also attention model obtained using soft alignment matrix. The soft alignment model of NMT system done by two methods of fast-align. First, match score checking the probability of a line input word according to the output and fast-align provides the highest probability of prediction. Second, mass score probability provides the weight of all the probability mass assigned to the alignment in the fast-align [92].

With the help of the score to handle encoding and many-to-many alignments of NMT of fast-align on the parallel data obtained from the alignment model to assign the input and output of the NMT system, then attention model guide the alignment training by assigning better word arrangements under supervising word alignment model of training.

3.6.6 BEAM SEARCH

Beam search-size is a different method to predict the translation not only having the highest score also with the next possible highest score words. The task of decoding an entire sentence for translation requires the highest probability of machine translation. Where in the traditional method of translation the problem encountered by the heuristic search technique provides another subpart of possible translation.

There was a straight forward method of parameters scoring with a beam size of resulting translations (BLEU) [93], if the parameter was going to increase with parameter beam size. In this fashion NMT model decoding the sentence to predict the next output of the word, which was having the highest scoring of being size, but it also shows next to the highest scoring words in the list of partial translations. Generally, the team score of the problem has required to better optimal quality of its size, almost all the optimum quality would be in range, but sometimes quality may decrease with larger being size. The main reason for this less quality has shorter translations of large beams.

3.6.7 *NOISY DATA*

Compare between traditional SMT with NMT there is reliable of noisy data in knowledge base of corrupted data like as misaligned context, improper translation sentence, wrong language sentence, etc. By applying probability distribution on the translated phrases used to build up a model, which the noisy data could be reduced as minimum sum of errors [94].

In neural translation the machine follows shuffling of misaligned data in the network for training data produces the maximum probability fuzzy on prediction of right translation. Even with NMT, also be not robust on prediction of translation, when mismatching of balancing between target language with training data. Increasing the ratio of training data for the target language as well as source language was the base model, if inadequate data over base model reflect to inappropriate output translation.

3.7 MACHINE TRANSLATION ENGINES

There are many organization involved in translation, with brings more economical provisions to their business, such as tourism will develop when the communication involves easy accessible on distinct locations. Also, the best example of website translation also makes it more popular as dynamic trending. Here some popular firms were establishing their huge investment in the field of machine translation; those establishments and the best processing technology in detail were discussed [95].

3.7.1 GOOGLE NEURAL MACHINE TRANSLATION (GNMT) SYSTEM

A decade before Google announces the product translate which was worked bases on the key algorithm of Phrase-Based Machine Translation, which becomes a rapid trending of intelligence over the machine, that improves in the field of speech recognition as well as image processing. It achieves a tremendous goal of the digital world, where it needs to improve challenges in the field of machine intelligence [96]. Now in the trending of GNMT System, which has the new technique to implement the machine translation with high quality on translations advanced of recurrent NN mechanism with it. As of traditional Phrase-Based Machine Translation split up the input into chunks or large chunks, has been translated with independent of its connectivity. When a NMT existence makes the entire input words into a single unit for translation, this provides more advantage over the existing one [97].

In the GNMT adapted the mechanism of recurrent NN to perform encoder and decoder for translation on both directions called as bidirectional recurrent NN. While applying the NN approach the input data was segmented into various chunks or sub-words then it allows training the model with huge amount datasets in a model called word piece model in the way of sequence conversion to sequence form. Compared to human translation with GNMT has made advanced development for translating a language pairs and also, it minimizes fault at the time of translation. In such reasons, a GNMT tries to occupy deeply various organizations, the current slogan of GNMT Google's corporation companies is "Do the right thing," because GNMT has the ability to solve advance things in tricky situations. It supports commercial promoting, management of bulk data set which were used by AI and machine learning applications and also used for different kinds of purposes (Figure 3.12).

In the following fantasy shows how the language Tamil was translated into English initially the network encoder has collect the letters and words with equal meaning of initial encoder and then forwarded to encoded face which begins to decode its internal meaning with different word alignment and then finally the encoder process the chunks of words translation will be mapped to the decoder, which is called target language English provides connectivity between number of chunks in encoder with a single decoder to make appropriate predictions. In the green line shows, connectivity between the chunks and the blue line shows the transparency of encoded-words and decoded words to get attention to frame into a final sentence. There are number of tool kits are available to do translation as a product based on numbers [98].

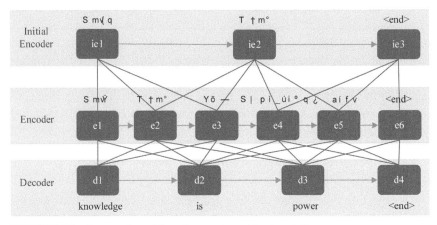

FIGURE 3.12   Progression of Google neural machine translation.

Three basic hitches:

1. Some calculation operations to build misapprehension of huge advancement;
2. By handling big data machine learning human translation cant able to handle huge data for language pair conversion; and
3. The improvement of GNMT over human translation was done in the reinforcement method of modifying with database.

There were number advantages over the multilingual translation of Google NMT system, in the form of simplicity, to use the single model for multiple language translation, when low resource available for translate between two languages then datasets for the different language will avail for translate in a better way with less data resource [99].

3.7.2   *OPEN NEURAL MACHINE TRANSLATION (NMT)*

Open NMT has open-source NMT, and it is often referred to as neural sequence modeling [100] (Figure 3.13).

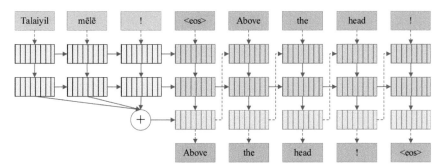

FIGURE 3.13   A sample open neural machine translation.

An Open NMT system is easy to use and the systems are also more proficient. The use of these systems can be started at the end of the year 2016. These open NMT systems are can be implemented and used by researchers, professionals, and scholastic in organizations.

Three core implementations of open NMT are:

- **Open NMT-Lua:** LuaTorch was urbanized by open MNT. It includes more variant features and a well-structured code.
- **OpenNMT-py:** The contemporary usage of OpenNMT-Lua is Open NMT-py(PyTorch). At first, it was designed and used by the Facebook AI team, for better utilization of predominant resources.
- **OpenNMT-tf:** The main focal point of TensorFlow is achieved better performance and most utilization of newly available resources.

Features contain:

- Establish easy communication between sources to destination.
- It follows well replicas and measures.
- Latest updating of characteristics leads to better understandability.
- It is capable to do different tasks like image recognition, voice recognition, speech recognition and conversions like image to text, text to image.
- Organizations and academicians can be vigorously used.

General Concerns of NMT:

Organizations are changing to NMT based machine translation systems because the translation can be fast and also automatic. HarvardNLP shows the current advances and efficient use of NLP and their applications like captioning of image, exchanging of ideas and so on. In deep learning NMT is one of the rapid growths in the branch of AI.

Training of NMT model contains two files or documents one for source and another one for target. By using the translation standards of WMT, translation is possible from source to target.

For training the systems the computers needed are:

- Small size data set model-contains GPU which support CUDA;
- Medium size data set model-needs at least 4GB computers;
- Full-size data set model-needs 8–12 GB computers.

Sometimes we can able to access pre-trained models. For such kind of standard free models needs less training. For acquiring translated data from source-to-target, have to give proper training for proper usage of data or files in OpenNMT and it uses parallel corpora open source technology [101].

An OpenNMT can work on other tribulations like summarization, dialogue, and tree-generation. OpenNMT is much efficient than neural systems like Neural Attention Model for Abstractive Summarization. For these kinds of situations, these systems contain miniature code for is effectual use of the applications. OpenNMT also used to implement image to sequence generation but it contains a little bit amount of excess code for translating im2text and so on [102].

3.7.3   DEEP NEURAL MACHINE TRANSLATION (DEEPNMT)

Deep neural machine translation (DeepNMT) is the extension of NMT. This technology is always updating over several years. At the evolution days of machine translation technology, one of the most dreadful tasks is converting English to other languages (like Asian languages). Philipp Koehn initiates the concept of SMT and fashioned a huge corpus of bilingual content derived from European Parliament documents and defined its first version of SMT. Later Philipp worked on in-depth knowledge of MT with the help of Chief Scientists [103].

With the increasing growth of AI for translating or differentiating between various languages in GNMT takes place a crucial role for translation purpose. And also, it uses different schemas for translation purposes. One of the ongoing machine translation techniques is DeepNMT and it is recommended in commercial applications. Every technique is mainly focused on quality factors for those we have to consider their countermeasures and issues related to it.

3.7.3.1 MACHINE TRANSLATION QUALITY

BLEU metric is another one now most of the human translators 15 out of 20 are using. It is a new neural machine provider that people are using it tries to clear the distinction between the machine and human translators. For dealing with such kind of issues are, first identify the linguist which is important to improve translation quality. We have to identify the MT provider for increasing the metric confidence. The second issue has to identify the size and length of the sentence for identifying the rationale of language level quality. By using DeepL the translators are trying to distinguish these kinds of issues raised in the machine to human translation [104].

By using the full list of criterion BLEU, we need to establish to endow with a solid and consistent metric. A BLEU is to improve the quality of machine translation; sometimes it is similar or less compared to SMT and RBMT. BLEU is unskilled and unacceptable when the metric is inappropriate.

NMT cannot able to handle if the test data is out of bound in such situations we are using SMT, for handling such things we have to train MT engine for handling the test data, if we can have partiality on test data its comfortable to train well. For suppose if you are searching some data in Google have to identify what kind of data it is and check whether any similar meanings are there for recalling such words. With the help of trained data, only it is possible to identify such cases [105].

3.7.3.2 MT PROVIDER PURPOSE

Deep Learning tries to translate whatever for anybody, any stage in current day's Google translator and Microsoft translator are the best translators for machine translation systems. We can able to translate whatsoever the data by using a machine translation engine. Sometimes writing a sentence creates different meanings, let us consider the following:

Sentence- "I went to the bank"

Different ways of writing:

- The water was icy. "*I went to the bank.*" My feet ruined into the mire.

- I have no money. *"I went to the bank."* But the bank is closed.
- I was at the top of the last round and the final turn. *"I went to the bank."* The g-force dragged me down into the seat.

For fast understanding and translation purposes, MT can divide the given sentence, for suppose other supplementary sentences related to "I went to the bank" are it may be river bank or ATM or bank where the situation we are using the meaning can vary. In general, the system understanding the exact meaning of a sentence is a bit complex task for MT providers [106].

Customized MT engines are considered for MT providers to determine the concerns related to sentences or words which indicate the same meaning but the usage of the situation is different. Consider one more example "VIRUS" is related to computer miscellaneous or affected to humans or affected to other living beings or trees. Because of these, kind of mixed meaning training MT is compulsorily needed. Omniscience Technologies has developed a variety of customized tools in this space that forms both bilingual and monolingual data. Delaying the translation may degrade the translation and speed of production. To overcome those have to use skillfully trained MT engines to increase the quality of translation.

Routing SMT via NMT via DeepNMT: Different considerations show that to traverse SMT to NMT to DeepNMT. With the increase of MT technology, NMT has capable to do work within a day and speed up translation by using single GPU it translates 5000 to 50,000 words per minute. The advancements of NMT technology towards commercial use will occupy different areas and training also needed for completing the task faster and decreasing complexity. Deep learning uses different algorithms for executing the longer sequence of instructions. It creates a neural layer for computing the complex tasks [107]. And use encoder and decoder technique for understanding input data and to process data. A recent comparison of NMT and DeepNMT is language pairs (English > Tamil and Tamil > English) it needs a trained set for evaluating identical data also to consider different metrics and specialized engines (Table 3.1).

The above metrics demonstrate the following; the data has to be Google or in Bing for handling different metrics served by a machine translation engine. Changing of NMT to Deep NMT is prominent in technology growth and the BLEU rate also varies. For training indistinguishable data, language pair needs training for translating EN-TA, TA-EN. The technology advancement taking place in the market is MT to Deep NMT or deep learning. Generic MT is a specialized customized engine to train the data sets. Other engines like OpenNMT produce optimal quality. All data in MT cannot produce eminence have to focus on what the customer needs an optimal solution.

TABLE 3.1 Different Metrics Served by Machine Translation Engine

Language Pair	MT Engine	BLEU	F-Measure	Levenshtein Distance	TER
EN-TA	Omniscient Deep NMT	46.03	75	12.65	33.94
	OmniSci NMT	34.75	68	17.52	46.12
	Google	29.87	64	18.64	50.03
	Bing	21.01	58	22.73	59.47
TA-EN	Omniscien Deep NMT	34.23	72	24.56	42.15
	OmniSci NMT	26.46	65	26.21	50.46
	Google	21.43	61	30.20	54.32
	Bing	16.05	55	37.46	66.51

3.7.4 PURE NEURAL MACHINE TRANSLATION (NMT)

One of the recent steps forward in the field of AI is deep learning. Sometimes deep learning is also calling as self-learning. While we are watching some movies, in those movies machines are more elegant compared to human's fine thought in the increase of technology. Traditional systems contain some input data embedded in the system depends on the input data the data processing can take place. But in the increasing growth of technology, the systems have the ability to think and act like humans a robotic technology is the best example for these kinds of systems [108].

A SYSTRAN system was first developed PNMT engines. Artificial NN uses contemporary engines like SMT and RBMT or a hybrid (combination of SMT and RBMT) approaches. In these approaches, we are using deep learning complex algorithms to increase translation quality because of this training process the systems has the ability to act like humans.

Previously the translation takes place word-by-word or unit by unit. For increasing the speed and accuracy of translation, some automated translation software's has been developed. SYSTRAN introduces latest PNMT approach for translating whole sentence or block of words for a particular topic [109].

An artificial neural network uses contemporary engines like SMT and RBMT or a hybrid (Combination of SMT and RBMT) approaches. In these approaches we are using deep learning complex algorithms to increase translation quality because of this training process the systems has the ability to act like humans.

For the translating, the machine translation output the main objective is to find out the consistency in data, repetitions in data and data reliability. BLEU machine is not always consistent it also contains some faults for handling some human assessments. Misunderstanding or mis-manipulation of source and target language leads to such inconsistencies. The rating process for BLEU model for language pairs decides exactness of NMTS method of human translators. With the help of SYSTRAN Google engine, the translation results will be consistent.

- For massive handling of resources NMT is used determination machine translation quality.
- NMT difficulties (training and implication speediness, vocabulary difficulty, misplaced words, etc.) will be resolved rather than later.
- At the time of segment translation some MT looks like human translation.
- Adaptive machine translation become more proficient translation in production systems like Facebook, Microsoft, Google, Baidu, and Systran because NMT adapts more rapidly for handling counteractive responses.

Specialized Pure Neural MT contains specialization which means post-training process that every SYSTRAN pure NMT machines are using for training the systems like humans which can work efficiently and also the systems developed as per user-specific dominion. PNMT contains more specialized and powerful well-trained algorithms a SYSTRAN is one of the good tool for pure NMT. It translates documents are of the identical or more quality than human-translated contented. In earlier engines, large quantity of data is compulsory to enrich its potential. But in current artificial neural network SYSTRAN systems has their own limitations to overcome the weakness which occurred in earlier engines and improves the capacity of translation [110] (Figure 3.14).

In past neural network engines, a large amount of data can't able to translate. To overcome this week point artificial neural network uses a SYSTRAN engine for developing the translation capability with the combination of previous technology with the present technology.

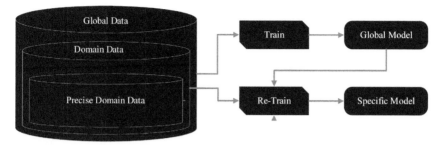

FIGURE 3.14 Pure NMT (train with global and re-train with precise data).

1. **Data Security:** For providing data confidentiality or data security or data protection, the enterprise PNMT is installing their own firewalled network. For preserving the data sometimes, the internet connectivity is also not needed. With the help of SYSTRAN, the things are to be kept as secure [111].

2. **SYSTRAN Enterprise Includes PNMT:** The PNMT with SYSTRAN edition uses different translation engines. Let's have a look each of them:
   • Rule-based assures translation reliable, more usability in different aspects and better performance considerations.
   • Statistical components use corpora for better usage of unilingual and multilingual languages which are helpful for making translation excellence and cost reduction for a specified province.
   • Hybrid it unites rule-based and statistical into a single technology and generates an engine for providing good accuracy.
   • PNMT uses different companies for delivering standard specific data.

3.8 CONCLUSION AND FUTURES OF MACHINE TRANSLATION

Following the machine translation of online learning with sequence of training for post-editing method includes three phases of prediction approach, correct answer approach, parameter update approach. In post-editing phases of work, the system produces the correct translation from source input given to the system applied on the training dataset with an updated method [112]. However, the approach of post-editing was utilized by the translation system [113]. The reinforcement method applied for new source language translation to extract target language, acquired from various nodes of data as compound training data in adaptive machine translation in order to overcome mistakes should not occurs repeatedly [114].

NMT is the trending translation paradigm of machine translation, which was overcome for some extent from the existing machine translations approach. As compared with other translation approaches NMT does not follow the mathematical procedure for best prediction of outputs, instead of that neuron of nodes applied to artificial neutral networks apprehend with many steps like training, encoding, decoding, attention, etc. From this above discussion makes the different computational processes for NMT that output produces will compare to the other existing translation approaches.

KEYWORDS

- **deep learning**
- **machine learning**
- **machine translation**
- **neural machine translation**
- **neural network**
- **statistical machine translation**

REFERENCES

1. Somers, H., (2003). *"Computers and Translation: A Translator's Guide"* (Vol. 35). John Benjamins Publishing.
2. Bahdanau, D., Kyunghyun, C., & Yoshua, B., (2014). *"Neural Machine Translation by Jointly Learning to Align and Translate."* arXiv preprint arXiv:1409.0473.
3. Galley, M., Mark, H., Kevin, K., & Daniel, M., (2004). *"What's in a Translation Rule."* Columbia Univ. New York Dept of Computer Science.
4. Castilho, S., Joss, M., Federico, G., Iacer, C., John, T., & Andy, W., (2017). "Is neural machine translation the new state of the art?" *The Prague Bulletin of Mathematical Linguistics, 108*(1), 109–120.

5. Bojar, O., Rajen, C., Christian, F., Yvette, G., Barry, H., Matthias, H., Antonio, J. Y., et al., (2016). "Findings of the 2016 conference on machine translation." In: *ACL 2016 First Conference on Machine Translation (WMT16)* (pp. 131–198). The Association for Computational Linguistics.

6. Brown, P. F., John, C., Stephen, A. D. P., Vincent, J. D. P., Frederick, J., Jennifer, C. L., & Robert, L. M., (1998). *"System for Parametric Text to Text Language Translation."* U.S. Patent 5,805,832.

7. Meermans, D. R., (1998). *"Automatic Voice/Text Translation of Phone Mail Messages."* U.S. Patent 5,712,901.

8. Dutoit, T., (1999). *"A Short Introduction to Text-to-Speech Synthesis."*

9. Graves, A., Santiago, F., Faustino, G., & Jürgen, S., (2006). "Connectionist temporal classification: Labeling unsegmented sequence data with recurrent neural networks." In: *Proceedings of the 23rd International Conference on Machine Learning* (pp. 369–376). ACM.

10. Graves, A., Abdel-rahman, M., & Geoffrey, H., (2013). "Speech recognition with deep recurrent neural networks." In: *2013 IEEE International Conference on Acoustics, Speech and Signal Processing* (pp. 6645–6649). IEEE.

11. Mikolov, T., Wen-Tau, Y., & Geoffrey, Z., (2013). "Linguistic regularities in continuous space word representations." In: *Proceedings of the 2013 Conference of the North American Chapter of the Association for Computational Linguistics: Human Language Technologies* (pp. 746–751).

12. Waibel, A., Ajay, N. J., Arthur, E. McNair, Hiroaki, S., Alexander, G. H., & Joe, T., (1991). "JANUS: A speech-to-speech translation system using connectionist and symbolic processing strategies." In: *Proceedings. ICASSP 91: 1991 International Conference on Acoustics, Speech, and Signal Processing* (pp. 793–796). IEEE.

13. Luong, M. T., & Christopher, D. M., (2015). "Stanford neural machine translation systems for spoken language domains." In: *Proceedings of the International Workshop on Spoken Language Translation* (pp. 76–79).

14. Du, Y., Wei, W., & Liang, W., (2015). "Hierarchical recurrent neural network for skeleton based action recognition." In: *Proceedings of the IEEE Conference on Computer Vision and Pattern Recognition* (pp. 1110–1118).

15. Hutchins, W. J., (1995). "Machine translation: A brief history." In: *Concise History of the Language Sciences* (pp. 431–445). Pergamon.

16. Callison-Burch, C., Philipp, K., & Miles, O., (2006). "Improved statistical machine translation using paraphrases." In: *Proceedings of the Main Conference on Human Language Technology Conference of the North American Chapter of the Association of Computational Linguistics* (pp. 17–24). Association for Computational Linguistics.

17. Graves, A., & Jürgen, S., (2005). "Frame wise phoneme classification with bidirectional LSTM and other neural network architectures." *Neural Networks, 18*(5/6), 602–610.

18. Hutchins, J., (2005). "Current commercial machine translation systems and computer-based translation tools: System types and their uses." *International Journal of Translation, 17*(1/2), 5–38.

19. Bharati, A., Vineet, C., Rajeev, S., & Ramakrishnamacharyulu, K. V., (1995). *"Natural Language Processing: A Paninian Perspective."* New Delhi: Prentice-Hall of India.

20. Graves, A., & Navdeep, J., (2014). "Towards end-to-end speech recognition with recurrent neural networks." In: *International Conference on Machine Learning* (pp. 1764–1772).

21. Koehn, P., (2009). *"Statistical Machine Translation."* Cambridge University Press.

22. Hutchins, W. J., & Harold, L. S., (1992). *"An Introduction to Machine Translation"* (Vol. 362). London: Academic Press.

23. Tu, Z., Zhengdong, L., Yang, L., Xiaohua, L., & Hang, L., (2016). *"Modeling Coverage for Neural Machine Translation.* "arXiv preprint arXiv:1601.04811.

24. Manaris, B., (1998). "Natural language processing: A human-computer interaction perspective." In: *Advances in Computers* (Vol. 47, pp. 1–66). Elsevier.

25. Och, F. J., (2003). "Minimum error rate training in statistical machine translation." In: *Proceedings of the 41st Annual Meeting on Association for Computational Linguistics* (Vol. 1, pp. 160–167). Association for Computational Linguistics.

26. Okamoto, E., & Atsushi, O., (1987). *"Multiple-Parts-of-Speech Disambiguating Method and Apparatus for Machine Translation System."* U.S. Patent 4,661,924.

27. Marcu, D., William, W., & Felix, L., (2014). *"Customizable Machine Translation Service."* U.S. Patent 8,831,928.

28. Kumano, A., Hiroyasu, N., Seiji, M., Hisahiro, A., & Shin-Ya, A., (1992). *"Machine Translation System."* U.S. Patent 5,091,876.

29. Richardson, S., & Richard, R., (2008). *"Adaptive Machine Translation Service."* U.S. Patent 7,383,542.

30. Kaji, H., & Yoshihiko, N., (1986). *"Displaying and Correcting Method for Machine Translation System."* U.S. Patent 4,599,612.

31. Gao, J., & Jie, Y., (2001). "An adaptive algorithm for text detection from natural scenes." In: *Proceedings of the 2001 IEEE Computer Society Conference on Computer Vision and Pattern Recognition. CVPR 2001* (Vol. 2, pp. II–II). IEEE.

32. Reiss, K., (2014). *"Translation Criticism-Potentials and Limitations: Categories and Criteria for Translation Quality Assessment."* Routledge.

33. Brown, P. F., John, C., Stephen, A. D. P., Vincent, J. D. P., Fredrick, J., John, D. L., Robert, L. M., & Paul, S. R., (1990). "A statistical approach to machine translation." *Computational Linguistics, 16*(2).

34. Simard, M., Nicola, U., Pierre, I., & Roland, K., (2007). "Rule-based translation with statistical phrase-based post-editing." In: *Proceedings of the Second Workshop on Statistical Machine Translation* (pp. 203–206). Association for Computational Linguistics.

35. Hull, D. A., & Gregory, G., (1996). "Querying across languages: A dictionary-based approach to multilingual information retrieval." In: *Proceedings of the 19th Annual International ACM SIGIR Conference on Research and Development in Information Retrieval* (pp. 49–57). ACM.

36. Grefenstette, G., (1999). "The World Wide Web as a resource for example-based machine translation tasks." In: *Proceedings of the ASLIB Conference on Translating and the Computer* (Vol. 21).

37. Llitjós, A. F., Jaime, G. C., & Alon, L., (2005). *"A Framework for Interactive and Automatic Refinement of Transfer-Based Machine Translation."*

38. Mitamura, T., Eric, N., & Jaime, G. C., (1991). *"An Efficient Interlingua Translation System for Multi-Lingual Document Production."*

39. Grefenstette, G., (1999). "The World Wide Web as a resource for example-based machine translation tasks." In: *Proceedings of the ASLIB Conference on Translating and the Computer* (Vol. 21).

40. Somers, H., (1999). "Example-based machine translation." *Machine Translation, 14*(2), 113–157.

41. Luong, M. T., & Christopher, D. M., (2016). *"Achieving Open Vocabulary Neural Machine Translation With Hybrid Word-Character Models."* arXiv preprint arXiv: 1604.00788.

42. Habash, N., (2004). "The use of a structural n-gram language model in generation-heavy hybrid machine translation." In: *Natural Language Generation* (pp. 61–69). Springer, Berlin, Heidelberg.

43. Koehn, P., (2004). "Statistical significance tests for machine translation evaluation." In: *Proceedings of the 2004 Conference on Empirical Methods in Natural Language Processing*.

44. Chung, J., Caglar, G., Kyung, H. C., & Yoshua, B., (2014). *"Empirical Evaluation of Gated Recurrent Neural Networks on Sequence Modeling."* arXiv preprint arXiv: 1412.3555.

45. Luong, M. T., Hieu, P., & Christopher, D. M., (2015). *"Effective Approaches to Attention-Based Neural Machine Translation."* arXiv preprint arXiv:1508.04025.

46. Jean, S., Kyunghyun, C., Roland, M., & Yoshua, B., (2014). *"On Using Very Large Target Vocabulary for Neural Machine Translation."* arXiv preprint arXiv:1412.2007.

47. Schuster, M., & Kuldip, K. P., (1997). "Bidirectional recurrent neural networks." *IEEE Transactions on Signal Processing, 45*(11), 2673–2681.

48. Sennrich, R., Barry, H., & Alexandra, B., (2015). *"Improving Neural Machine Translation Models With Monolingual Data."* arXiv preprint arXiv:1511.06709.

49. Zhang, J., & Chengqing, Z., (2015). *"Deep Neural Networks in Machine Translation: An Overview."*

50. Kalchbrenner, N., Lasse, E., Karen, S., Aaron Van Den, O., Alex, G., & Koray, K., (2016). *"Neural Machine Translation in Linear Time."* arXiv preprint arXiv:1610.10099.

51. Bengio, Y., Réjean, D., Pascal, V., & Christian, J., (2003). "A neural probabilistic language model." *Journal of Machine Learning Research, 3*, 1137–1155.

52. Goldberg, Y., (2017). "Neural network methods for natural language processing." *Synthesis Lectures on Human Language Technologies, 10*(1), 1–309.

53. Sutskever, I., Oriol, V., & Quoc, V. L., (2014). "Sequence to sequence learning with neural networks." In: *Advances in Neural Information Processing Systems* (pp. 3104–3112).

54. Devlin, J., Rabih, Z., Zhongqiang, H., Thomas, L., Richard, S., & John, M., (2014). "Fast and robust neural network joint models for statistical machine translation." In: *Proceedings of the 52nd Annual Meeting of the Association for Computational Linguistics* (Vol. 1, pp. 1370–1380).

55. Luong, M. T., Hieu, P., & Christopher, D. M., (2015). *"Effective Approaches to Attention-Based Neural Machine Translation."* arXiv preprint arXiv:1508.04025.

56. Bentivogli, L., Arianna, B., Mauro, C., & Marcello, F., (2016). *"Neural versus Phrase-Based Machine Translation Quality: A Case Study."* arXiv preprint arXiv:1608.04631.

57. Jean, S., Kyunghyun, C., Roland, M., & Yoshua, B., (2014). *"On Using Very Large Target Vocabulary for Neural Machine Translation."* arXiv preprint arXiv:1412.2007.

58. Tu, Z., Yang, L., Lifeng, S., Xiaohua, L., & Hang, L., (2017). "Neural machine transla-tion with reconstruction." In: *Thirty-First AAAI Conference on Artificial Intelligence.*

59. Gulcehre, C., Orhan, F., Kelvin, X., Kyunghyun, C., Loic, B., Huei-Chi, L., Fethi, B., Holger, S., & Yoshua, B., (2015). *"On Using Monolingual Corpora in Neural Machine Translation."* arXiv preprint arXiv:1503.03535.

60. Cho, K., Bart, V. M., Caglar, G., Dzmitry, B., Fethi, B., Holger, S., & Yoshua, B., (2014). *"Learning Phrase Representations Using RNN Encoder-Decoder for Statistical Machine Translation."* arXiv preprint arXiv:1406.1078.

61. Cho, K., Bart, V. M., Dzmitry, B., & Yoshua, B., (2014). *"On the Properties of Neural Machine Translation: Encoder-Decoder Approaches."* arXiv preprint arXiv:1409.1259.

62. Meng, F., Zhengdong, L., Mingxuan, W., Hang, L., Wenbin, J., & Qun, L., (2015). *"Encoding Source Language with Convolutional Neural Network for Machine Translation."* arXiv preprint arXiv:1503.01838.

63. Chung, J., Kyunghyun, C., & Yoshua, B., (2016). *"A Character-Level Decoder Without Explicit Segmentation for Neural Machine Translation."* arXiv preprint arXiv:1603.06147.

64. Cho, K., Bart, V. M., Dzmitry, B., & Yoshua, B., (2014). *"On the Properties of Neural Machine Translation: Encoder-Decoder Approaches."* arXiv preprint arXiv: 1409.1259.

65. LeCun, Y., Yoshua, B., & Geoffrey, H., (2015). "Deep learning." *Nature, 521*(7553), 436.

66. Schmidhuber, J., (2015). "Deep learning in neural networks: An overview." *Neural Networks, 61*, 85–117.

67. Deng, L., & Dong, Y., (2014). "Deep learning: Methods and applications." *Founda-tions and Trends® in Signal Processing, 7*(3/4), 197–387.

68. Bengio, Y., Aaron, C., & Pascal, V., (2013). "Representation learning: A review and new perspectives." *IEEE Transactions on Pattern Analysis and Machine Intelligence, 35*(8), 1798–1828.

69. Radford, A., Luke, M., & Soumith, C., (2015). *"Unsupervised Representation Learn-ing with Deep Convolutional Generative Adversarial Networks."* arXiv preprint arXiv:1511.06434.

70. Zaremba, W., Ilya, S., & Oriol, V., (2014). *"Recurrent Neural Network Regularization."* arXiv preprint arXiv:1409.2329.

71. Kalchbrenner, N., & Phil, B., (2013). "Recurrent continuous translation models." In: *Proceedings of the 2013 Conference on Empirical Methods in Natural Language Processing* (pp. 1700–1709).

72. Munday, J., (2012). *Evaluation in Translation: Critical Points of Translator Decision-Making.* Routledge.

73. Isozaki, H., Tsutomu, H., Kevin, D., Katsuhito, S., & Hajime, T., (2010). "Automatic evaluation of translation quality for distant language pairs." In: *Proceedings of the 2010 Conference on Empirical Methods in Natural Language Processing* (pp. 944–952). Association for Computational Linguistics.

74. Bangalore, B., German, B., & Giuseppe, R., (2001). "Computing consensus transla-tion from multiple machine translation systems." In: *IEEE Workshop on Automatic Speech Recognition and Understanding* (pp. 351–354). ASRU'01, IEEE.

75. Krings, H. P., (2001). *Repairing Texts: Empirical Investigations of Machine Translation Post-Editing Processes* (Vol. 5). Kent State University Press.

76. Pascanu, R., Tomas, M., & Yoshua, B., (2013). "On the difficulty of training recurrent neural networks." In: *International Conference on Machine Learning* (pp. 1310–1318).

77. Aziz, W., Sheila, C., & Lucia, S., (2012). "PET: A tool for post-editing and assessing machine translation." In: *LREC* (pp. 3982–3987).

78. Green, S., Jeffrey, H., & Christopher, D. M., (2013). "The efficacy of human post-editing for language translation." In: *Proceedings of the SIGCHI Conference on Human Factors in Computing Systems* (pp. 439–448). ACM.

79. Arenas, A. G., (2008). "Productivity and quality in the post-editing of outputs from translation memories and machine translation." *Localization Focus, 7*(1), 11–21.

80. Costa-Jussa, M., R., & José, A. R. F., (2016). *"Character-Based Neural Machine Translation."* arXiv preprint arXiv:1603.00810.

81. Artetxe, M., Gorka, L., Eneko, A., & Kyunghyun, C., (2017). *"Unsupervised Neural Machine Translation."* arXiv preprint arXiv:1710.11041.

82. Koehn, P., & Rebecca, K., (2017). *"Six Challenges for Neural Machine Translation."* arXiv preprint arXiv:1706.03872.

83. Galley, M., Mark, H., Kevin, K., & Daniel, M., (2004). *What's in a Translation Rule?* Columbia University New York Dept of Computer Science.

84. Sennrich, R., Barry, H., & Alexandra, B., (2015). *"Improving Neural Machine Translation Models with Monolingual Data."* arXiv preprint arXiv:1511.06709.

85. Luong, Minh-Thang, Ilya, S., Quoc, V. L., Oriol, V., & Wojciech, Z., (2014). *"Addressing the Rare Word Problem in Neural Machine Translation."* arXiv preprint arXiv:1410.8206.

86. Luong, M. T., Ilya, S., Quoc, V. L., Oriol, V., & Wojciech, Z., (2014). *"Addressing the Rare Word Problem in Neural Machine Translation."* arXiv preprint arXiv:1410.8206.

87. Sennrich, R., Barry, H., & Alexandra, B., (2015). *"Neural Machine Translation of Rare Words with Sub Word Units."* arXiv preprint arXiv:1508.07909.

88. Sennrich, R., Barry, H., & Alexandra, B., (2015). *"Neural Machine Translation of Rare Words with Sub Word Units."* arXiv preprint arXiv:1508.07909.

89. Pouget-Abadie, J., Dzmitry, B., Bart, V. M., Kyunghyun, C., & Yoshua, B., (2014). *"Overcoming the Curse of Sentence Length for Neural Machine Translation Using Automatic Segmentation."* arXiv preprint arXiv:1409.1257.

90. Johnson, M., Mike, S., Quoc, V. L., Maxim, K., Yonghui, W., Zhifeng, C., Nikhil, T., et al., (2017). "Google's multilingual neural machine translation system: Enabling zero-shot translation." *Transactions of the Association for Computational Linguistics, 5*, 339–351.

91. Bahdanau, D., Kyunghyun, C., & Yoshua, B., (2014). *"Neural Machine Translation by Jointly Learning to Align and Translate."* arXiv preprint arXiv:1409.0473.

92. Liu, Y., & Maosong, S., (2015). "Contrastive unsupervised word alignment with non-local features." In: *Twenty-Ninth AAAI Conference on Artificial Intelligence.*

93. Papineni, K., Salim, R., Todd, W., & Wei-Jing, Z., (2002). "BLEU: A method for automatic evaluation of machine translation." In: *Proceedings of the 40th Annual Meeting on Association for Computational Linguistics* (pp. 311–318). Association for Computational Linguistics.

94. Vaswani, A., Yinggong, Z., Victoria, F., & David, C., (2013). "Decoding with large-scale neural language models improves translation." In: *Proceedings of the 2013 Conference on Empirical Methods in Natural Language Processing* (pp. 1387–1392).
95. Junczys-Dowmunt, M., Tomasz, D., & Hieu, H., (2016). *"Is Neural Machine Translation Ready for Deployment? A Case Study on 30 Translation Directions."* arXiv preprint arXiv:1610.01108.
96. Wu, Yonghui, Mike, S., Zhifeng, C., Quoc, V. L., Mohammad, N., Wolfgang, M., Maxim, K., et al., (2016). *"Google's Neural Machine Translation System: Bridging the Gap Between Human and Machine Translation."* arXiv preprint arXiv:1609.08144.
97. Johnson, M., Mike, S., Quoc, V. L., Maxim, K., Yonghui, W., Zhifeng, C., Nikhil, T., et al., (2017). "Google's multilingual neural machine translation system: Enabling zero-shot translation." *Transactions of the Association for Computational Linguistics*, 5339–5351.
98. Neubig, G., (2017). *"Neural Machine Translation and Sequence-to-Sequence Models: A Tutorial."* arXiv preprint arXiv:1703.01619.
99. Le, Q. V., & Mike, S., (2016). *"A Neural Network for Machine Translation, at Production Scale."* Google Research Blog.
100. Klein, G., Yoon, K., Yuntian, D., Jean, S., & Alexander, M. R., (2017). *"Open NMT: Open-Source Toolkit for Neural Machine Translation."* arXiv preprint arXiv: 1701.02810.
101. Britz, D., Anna, G., Minh-Thang, L., & Quoc, L., (2017). *"Massive Exploration of Neural Machine Translation Architectures."* arXiv preprint arXiv:1703.03906.
102. Tokui, S., Kenta, O., Shohei, H., & Justin, C., (2015). "Chainer: A next-generation open source framework for deep learning." In: *Proceedings of Workshop on Machine Learning Systems (LearningSys) in the Twenty-Ninth Annual Conference on Neural Information Processing Systems (NIPS)* (Vol. 5, pp. 1–6).
103. LeCun, Y., Yoshua, B., & Geoffrey, H., (2015). "Deep learning." *Nature, 521*(7553), 436.
104. Courbariaux, M., Yoshua, B., & Jean-Pierre, D., (2015). "Binary connect: Training deep neural networks with binary weights during propagations." In: *Advances in Neural Information Processing Systems* (pp. 3123–3131).
105. Collobert, R., & Jason, W., (2008). "A unified architecture for natural language processing: Deep neural networks with multitask learning." In: *Proceedings of the 25th International Conference on Machine Learning* (pp. 160–167). ACM.
106. Venugopalan, S., Huijuan, X., Jeff, D., Marcus, R., Raymond, M., & Kate, S., (2014). *"Translating Videos to Natural Language Using Deep Recurrent Neural Networks."* arXiv preprint arXiv:1412.4729.
107. Courbariaux, M., Itay, H., Daniel, S., Ran, El-Yaniv, & Yoshua, B., (2016). *"Binarized Neural Networks: Training Deep Neural Networks with Weights and Activations Constrained to +1 or –1."* arXiv preprint arXiv:1602.02830.
108. Crego, J., Jungi, K., Guillaume, K., Anabel, R., Kathy, Y., Jean, S., Egor, A., et al., (2016). *"Systran's Pure Neural Machine Translation Systems."* arXiv preprint arXiv:1610.05540.
109. Crego, J., Jungi, K., Guillaume, K., Anabel, R., Kathy, Y., Jean, S., Egor, A., et al., (2016). *"Systran's Pure Neural Machine Translation Systems."* arXiv preprint arXiv:1610.05540.

110. Dorr, B. J., Pamela, W. J., & John, W. B., (1999). "A survey of current paradigms in machine translation." In: *Advances in Computers* (Vol. 49, pp. 1–68). Elsevier.
111. Hermann, K. M., Tomas, K., Edward, G., Lasse, E., Will, K., Mustafa, S., & Phil, B., (2015). "Teaching machines to read and comprehend." In: *Advances in Neural Information Processing Systems* (pp. 1693–1701).
112. Hutchins, W. J., (1986). *Machine Translation: Past, Present, Future.* Springer.
113. Hutchins, W. J., (2001). "Machine translation over fifty years." *Histoire Epistémologie Langage, 23*(1), 7–31.
114. Slocum, J., (1985). "A survey of machine translation: Its history, current status, and future prospects." *Computational Linguistics, 11*(1), 1–17.

CHAPTER 4

Role of Machine Learning and Application Towards Information Retrieval in Cloud

MISHRA SAMBIT KUMAR,[1] MISHRA BROJO KISHORE,[2] and PRASAD SUMAN SOURAV[3]

[1]*Department of Computer Science and Engineering, Gandhi Institute for Education and Technology, Bhubaneswar, Odisha, India, E-mail: sambitmishra@gietbbsr.com*

[2]*Department of Computer Science and Engineering, GIET University, Gunupur, Odisha, India, E-mail: brojokishoremishra@gmail.com*

[3]*Department of MCA, Ajay Binay Institute of Technology, Cuttack, Odisha, India, E-mail: prasadsuman800@rediffmail.com*

ABSTRACT

In a broad sense, natural language processing (NLP) may be defined as the capacity of a set of computer instructions to be familiar with human interpretation language as a part of artificial intelligence (AI). In general, it may be concerned more precisely towards coding to process and analyze a huge amount of data. Somehow, Challenges in NLP may involve with speech recognition, understanding natural language as well as generating natural language. In the present scenario, it has been more focused on in supervised and semi-supervised learning techniques. In such cases, the algorithms associated with such techniques may be able to learn from data using a combination of annotated and non-annotated data. Practically, it may be too difficult and may produce less accurate results for a given amount of input data.

The primary intention of NLP may be linked to accomplish human-like language processing. Choosing the word during performance may be very deliberate, and may not be replaced with actual understanding. The complete perception of NLP may be associated with the following:

1. Making phrase and outlined the input text;
2. Conversion of the text into similar language in different form;
3. Queries about the text;
4. Inference from the text.

Considering the levels, it has been presumed that the levels of human language processing may be assigned in sequential manner and seems to be more dynamic. Accordingly, the information gained may also assist in lower level of analysis.

4.1   INTRODUCTION

Initially, the concept associated with the related content may be retrieved, observed, and linked with the similarities. After understanding the basic concepts, it may be required to be properly formulated. In the usual case, the contents may be associated with natural language. So in such a situation, text mining (TM) may be needed towards the retrieval of information along with some possible natural language processing (NLP) techniques. In this case, it may be primarily focused towards the retrieval of information from cloud linked with the virtual database, and it may be associated with retrieving entities, categorizing entities, clustering along with fact retrieval mechanisms. Cited as an example, towards accessing and retrieval of large volume of data and information, approaches may be made by the system to search its database linked to the user's query along with the relevancy. The retrieval mechanism of information may be segregated towards indexing, querying, comparison, as well as feed-back. The primary importance may be linked to accumulate and manage the relevant documents. In such cases, retrieval system may be adopted towards suitable representation of the documents as per requirements. In many cases, while employing the method of indexing to the descriptive queries, the response may be generated along with the indices and queries and may link to conceptual schema. In such situation, the complexity towards queries may arise along with limited scope, for example, indexing the system performance, generating the databases with interface. It is well understood that, in many situations, the user queries may initiate towards formation of a single key word or phrase. But, it may be difficult towards evaluation of indexing terms are used to describe the documents they want. Accordingly, the systems may employ simple query methods along with provision of facility towards users.

Again cited another example, in a query evaluation system, words along with indices may be matched with similar meaning. To acquire the mechanism, the retrieval systems sometimes may exploit the schema instead of reengineering the queries. So the system measurement may be done along with weighting terms related to query performance. After that, as per the conceptual schema, the information associated with the databases may be compared. In general, case, the learning algorithms (LAs) may be adopted to automate information retrieval (IR) processes, that may linked to minimize the inconsistency while classifying the documents. So it may be good enough to be familiar with rule induction, instance-based learning, neural networks (NN), genetic algorithms (GA), as well as analytic learning. In many applications, heuristics may be applied to generate structures towards representation of relationships implicit in the data. The analytic learning may be employed towards learn proofs or explanations for example situations using background knowledge. Towards achieving the large scale IR, it may be sometimes essential to keep track of background knowledge as well as complex structure associated with the analytic learning systems. For example, GAs may be least frequently applied towards instance-based learning (IBk) as well as analytical learning, as being inspired by biological genetic systems. In a similar fashion, decision tree (DT) and rule induction schemes may be treated as better developed machine learning techniques. In this case, the input data may be used towards the generation of explicit descriptions. According to the present scenario, the accuracy along with appropriateness may reflect towards the quality of the data supplied. In this regard, the algorithms have already been developed to select the most potent regularities towards the concept description, in which, if at all strong patterns appear by chance, or the data is irrelevant to the classification task, the concept description may be inadequate. In addition to that, different methodologies have also been applied towards determination of patterns suitable in the concept description. In many cases, divide-and-conquer algorithms may be quite suitable towards creation of protocols for each class. Sometimes the partial structured or semi-structured as well as diversity of natural data may be forced towards challenges to the application of machine learning. Also, this may cause problems towards satisfying a single query, and also may not directly compare ranking scores across collections linked with unique features.

In many cases, the necessity of machine learning may be considered towards specific areas like image or speech recognition, perception, etc. Accordingly, data may be provided towards the improvement of the performance, and by that, the extraction of relevant information through various techniques associated with data analysis may be quite feasible implementing machine learning towards social and economic relevancies.

4.2   LITERATURE SURVEY

Zakir et al. [1] in their work focused towards Apache project related to open source enterprise platform along with searching function. Accordingly, many synchronized tools were being accumulated in Apache Lucene's library to enhance and extend its full-text search capability. But, it was not an executable search application rather than a toolkit or IR system. In that scenario, it was merged and linked with associated techniques for index searching, TM along with IR.

Sanjay et al. [2] during their work have scaled out the file storage system for large distributed data-intensive applications. The basic idea, in this case, was to develop the distribute file System for web crawling application. The basic concepts associated with the application could be measured in terms of scalability, data mutability, communication protocol, replication strategy, and security.

Stephan Ewen et al. [3] in their work have discussed about 4th generation data processing engine along with special features. The data engine may be somehow having a general-purpose framework for large scale data analysis and it has been observed that the outcome is much better as compared to the previous one.

Matei et al. [4] in their work have focused towards the general-purpose framework and Resilient Distributed Datasets. The primary intention in this work may be to extend the basic services of the general-purpose framework towards memory computation techniques. In a simple term, it may be associated with a better solution. Practically it emphasizes the computational efficiency of iterative and recursive algorithms and interactive queries of data mining (DM).

Xin Liu et al. [5] in their study have focused towards the fundamental application of machine learning. In general, it aims towards recognizing complex patterns and generating intelligent decisions based on existing datasets. The main concept is to build systems as per human competence towards performing complex tasks.

James Surowiecki et al. [6] in their work have implemented the mechanisms of learning to transform the input of handwriting letters to the output of the standard recognized letter. As discussed, the principle of learning from data may be more equivalent to the mechanisms associated with trial-error as well as "The Wisdom of Crowds" and may be applied to reduce down to an acceptable level of convergence.

Paul et al. [7] in their work have elaborated on the cause linked to veracity dimension, connection to the quality and source issues of large scale initiative of data.

David Lazer et al. [8] in their work have predicted and identified issues related to Big Data hubris along with algorithm dynamics. Sometimes it may also traditional DM completely. The issue associated with algorithm dynamics may be responsible towards the improvement of the commercial services.

Bhat et al. [9] in their work focused towards machine learning techniques along with anomalies by splitting the task into service, user, host, and workflow. Each one may be implemented to analyze the behavior associated with the activities. In every instance, the time tracks also may allow for further classification of profiles towards detection of anomalous behaviors of the workflow.

Chase et al. [10] during their study have discussed the approaches towards reduction of overall energy consumption while maintaining quality of service as well as service level agreement. They also focused towards greedy algorithm along with dynamic mapping of energy consumption.

Tierney et al. [11] in their work have discussed about Language Integrated Query system which have been developed at Microsoft Research. It is nothing but general-purpose architecture for the execution of data-parallel applications.

Hido et al. [12] in their work have focused on online/real-time machine learning platform which was implemented on a distributed architecture. Accordingly, the system continuously was updated with each data sample and required no data storage and linked to classification problems.

Ghoting et al. [13] in their work have associated with NLP, DM along with advanced visualization capabilities. Practically, the author implemented a large amount of data in the linked environment with analytical tools.

Low et al. [14] in their work have discussed about business intelligence associated with a similar data warehouse. The cloud providers, in general, may focus on different features along with the predictive analysis framework.

Pednault et al. [15] in their work have discussed regarding cloud-based TM. Accordingly, the platform may support towards NLP capabilities along with TM platform. Also the concerned algorithm may use deep linguistic parsing, statistical NLP. In this case, the machine learning may be implemented to analyze semantic meta-data.

Eckerson et al. [16] in their work have discussed regarding basic Service providers, may be useful to achieve insights out of their large datasets. The users, in this case, may sometimes have new and unique anomalies, the possibility to perform fast, ad-hoc investigations on data.

Shiravi et al. [17] during the experimentation tried to develop machine learning-based security applications. The models implemented in the application may be associated with suitable experimentations along with a comprehensive set of intrusions. In general, different sets of features are extracted from these traces and are used to train the models.

Sommer et al. [18] in their work have focused towards security-related datasets. Accordingly, the machine learning models are being trained and evaluated over experimentally generated datasets.

Bhamare et al. [19] in their experimentation focused on the models linked with a particular dataset. This may be due to the fact that user packets may need to travel through different data centers/clouds, distributed across multiple locations operating in diversified environments.

Mchugh et al. [20] in their work have considered multiple datasets to provide the required comprehensiveness in testing machine learning models. Such testing of the learned models may provide a sense of robustness and applicability of the learned models.

4.3   APPLICATIONS OF MACHINE LEARNING

It may be very clear that the implementation mechanisms associated with machine learning may be responsible towards changes and reengineering the existing specific techniques. Accordingly, the applications towards machine learning may perform better towards specific tasks.

In many times, it may allow the system to learn directly from examples and experience in the form of data, as in this case, being associated with set of tasks along with large amount of data; tasks may be performed by detecting the patterns. In this manner, it may be a much better way to achieve the desired output. In a broad sense, it supports intelligent systems towards particular functions. It also allows the systems to perform specific tasks intelligently, by learning from examples and ignoring the pre-programmed rules.

In general, there are relevant mechanisms associated with machine learning as per following:

1. In supervised machine learning, the trained data may be labeled. Accordingly, each data may be categorized to more groups in different data points. The system may learn about structuring the trained data along with prediction mechanisms.
2. Unsupervised learning may be associated with learning without labels. Sometimes it may aim to detect the characteristics towards making the data points more or less similar to each other.
3. Reinforcement learning, in general, may tend towards unsupervised and supervised learning. In a typical reinforcement learning scheme, the specific entity may interact with its environment and may try to optimize. The primary intention in this case may be focused towards learning strategies.

As an example, machine learning techniques may be associated towards making decision towards diagnosis of symptoms and finding medicines. Preliminary it may allow the system to process written or verbal information towards extraction of information and implement the same towards treatment.

The application of machine learning with specific and large volume of training data may play vital role towards optimizing logistics and associated processes. It may be achieved redirecting towards storage facilities with retrieval mechanisms.

4.4 INFORMATION RETRIEVAL (IR)

In very clear term, the system associated with information storage and retrieval may make large volumes of text accessible to the users along with the approaches and goal of the system. The outlines related to the requirements may lead towards relevancy and occurrence of databases along with user's queries. Accordingly, the response to the query may be constructed using the indices and the operations provided by the system associated with conceptualization. It is really essential to model the complexity of the queries as per requirements. But, the users may not know about the indexing terms may be used to describe the databases.

4.4.1   TEXT SYSTEM MODELS ALONG WITH REPRESENTATION

While extracting information from the relevant databases by applying the specific retrieval mechanism, it may be essential to categorize the documents and generate relevancy with accuracy. In this situation, the natural language may have a specific myriad properties used to build the mechanisms towards the system. In fact, the speed of processing is an important factor in large-scale IR, so in reality, only a small set of easily extracted features may be used by any system.

4.4.2   INDEXING WITH SEMANTIC ANALYSIS

In many cases, there may be many ways to be associated with similar objects and it may not be necessary to have literals or words in common. Also, many words may mean quite variant things in different contexts associated with the problems. In fact, the representation may map the queries and documents into an array of significant factor weights leads to specific dimensionality of the new concept of array. The primary concern of the approach may be linked with semantic qualities. Accordingly, the query may be processed by representing its terms as a two-dimensional array in the space and then ranking documents by their proximity to the query terms. The same may be applied to compare the document clustering with query clustering.

4.4.3   SYSTEM ASSOCIATED WITH SEQUENTIAL APPROACH

Towards compressing the required text, many times the information may be grouped into strings of arbitrary length. In that scenario, to identify an index, each string may be used to be placed in a compressed version of data associated with the database. Again, to decode the data associated with the database, the hierarchy of the database may be considered and indices may be replaced with appropriate piece of text. In many cases, some specific techniques may be used towards grouping of characters linked to the basis of dividing the text into similar strings.

Now considering the logistic analysis which is nothing but an appropriate analysis towards dependent binary variables. Sometimes it may also be termed as predictive analysis. It may be implemented towards describing data and to explain the relationship between one dependent binary variable and one or more nominal, ordinal, interval or ratio-level independent variables.

4.5   STEPS TO MEASURE THE PERFORMANCE OF DATA ACQUISITION AND RETRIEVAL MECHANISM

- **Step 1:** Probabilistic measures may be predicted with binary values along with numerical and categorical.
- **Step 2:** Values linked to associated and acceptable range also may be considered ranging from 0 to 1.
- **Step 3:** Since the experiment may be considered with two possible values, the residuals may not be normally distributed about the predicted line.
- **Step 4:** Logistic analysis may be applied using maximum likelihood estimation technique to obtain the coefficients towards relating predictors.
- **Step 5:** Evaluation of statistical significance of every coefficient associated with the systems.
- **Step 6:** Adoption of testing mechanisms to assess the significance of prediction of each predictor.
- **Step 7:** Perform comparison of data along with retrieval mechanisms associated with the system.

4.6   APPLICATION OF METAHEURISTIC APPROACH TO THE SYSTEM

As being observed, the metaheuristic algorithms may be used for obtaining contemporary global optimization algorithms, computational intelligence (CI), and soft computing. These algorithms are in general nature-inspired with multiple interacting agents. The subset of metaheuristics is often termed as swarm intelligence algorithms, linked to swarm intelligence characteristics of biological agents. One of such approach may be termed as firefly algorithm. It is based on the flashing patterns and behavior of fireflies. The attractiveness maybe somehow proportional to the brightness. Accordingly, for any two flashing fireflies, the less bright one will move towards the brighter one. If there is no brighter one than a particular firefly, it will move randomly. Also, the brightness of a firefly may be obtained by the landscape of the objective function (Table 4.1).

TABLE 4.1   Data Servers with Associated Time of Acquisition

Sl. No.	Number of Data Servers	Percentage of Acquiring Data (%)	Acquisition Time Associated with Databases (ms)
1.	19	14	0.11
2.	24	19	0.14
3.	30	34	0.37
4.	40	69	0.77
5.	50	83	0.88

Considering Table 4.1, it has been observed that, the percentage of acquiring data may be directly proportional to the size of the server space. Also, the acquisition time associated with the database always depends on the allocation of data servers (Table 4.2).

TABLE 4.2   Data Centers with Associated Time of Acquisition

Sl. No.	Number of Data Centers	Percentage of Acquiring Data (%)	Acquisition Time Associated with Data Centers (ms)
1.	19	11	0.09
2.	30	16	0.24
3.	40	47	0.67
4.	50	69	0.88
5.	60	73	0.96

As reflected in Table 4.2, the acquisition time associated with the data centers depends on the allocation of data centers.

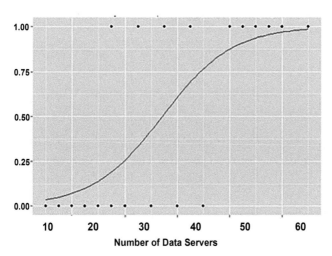

FIGURE 4.1   Probability of acquisition of data through database.

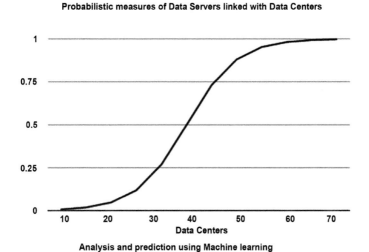

FIGURE 4.2   Probabilistic measures of data servers linked with data centers.

4.7   COMPLEXITY ASSOCIATED WITH THE SYSTEM

It is understood that the metaheuristic algorithms are simple and easy to implement. The main computational cost may be in the evaluations of objective functions. It has attracted much attention and been observed about the computation time for the digital image. It may also produce consistent and better performance in terms of time and optimality. Classifications and clustering are also another important areas of applications, and it has been observed that it may be efficiently used for clustering. Anyway, it may be associated with two distinct merits, i.e., automatic subdivision and ability of dealing with multimodality. The population may be automatically subdivided into subgroups, and each group may swarm around each mode or local optimum. Accordingly, the best global solution may be obtained. After that, the subdivision may permit the fireflies to be able to find all optima simultaneously if the population size may be sufficiently higher than the number of modes.

The virtualized data associated with the servers, in general, may be linked with computing infrastructure, data, and application services. The primary intention in such a case may be to build computing clusters and scale-out the computing capacity. Considering the large set of data in servers, the cost of processing along with computational capacity may be essential. In this regard, while processing the large dataset linked with subsequent models, the changes may be somehow required in the program states along with modularity. In general, being associated with machine learning, the generalized tasks may be synchronized and the workloads may be properly assigned. But somehow, it may not be so effective towards recursive processes.

4.8 IMPACT OF MACHINE LEARNING IN VIRTUAL DATABASE

In general, there are basic prerequisites towards virtual databases, i.e., low-cost resources and processing power. Initially, it may be associated with processing power with minimal cost and then the ability to process large size of data. In this regard, the impact of machine learning may be linked with cognitive computing. The storage amount of data in the virtual environment sometimes may be associated with machine learning process. Accordingly, it may provide applications in the virtual environment with sensor capabilities and also the applications may be able to perform the linked functions with suitable decisions. The integration of machine learning may enhance the requirement of virtual space. Also along with suitability of virtual servers, the enhancement of technology may be beneficial towards the relevant fields.

4.9 DISCUSSION AND FUTURE DIRECTION

Sometimes, the analysis to predict the requirement, as well as application for both tactical and strategic purposes, may be restricted because of prohibitive resource requirements. But the machine learning associated with the virtual environment may not be constraint towards all large sets of heterogeneous data or data enterprise. Also, the artificial intelligence (AI) systems may be performing better on virtual servers, as linked with low cost of operations, scalability, and huge processing power to analyze the huge amount of data. Accordingly, it may act as a source for the machine LAs. The algorithms associated with machine learning may gather information and perform better with suitability. AI research in a practical situation may result better in a virtual environment.

4.10 CONCLUSION

It has been observed that the machine learning technologies along with computation in virtual environment are emerging in the present days. In such a scenario, it may take some time to be fully functional and to be used in crucial sectors like healthcare, business, and banking. It is clear that machine learning may make the data easier to handle over the cloud. Along with repetitive AI research in cloud computing, it may become more and more intelligent. It may be stated that the supervised machine learning models will perform well with a particular dataset, and may perform satisfactorily with totally different datasets generated with different simulation or experimental conditions and environments.

KEYWORDS

- analytical learning
- clustering
- machine learning
- query terms
- supervised learning
- virtual database

REFERENCES

1. Dr. Zakir, L., & Abdulbasit, S., (2013). *Web Crawling and Data Mining with Apache Nutch.* Packt Publishing.
2. Sanjay, G., Howard, G., & Shun-Tak, L., (2003). *"The Google File System"* (pp. 1–15). SOSP'03.
3. Stephan, E., Sebastian, S., Kostas, T., Daniel, W., & Volker, M., (2013). Iterative parallel processing with stratosphere an inside look. *Proceedings of the 2013 ACM SIGMOD International Conference on Management of Data, SIGMOD 13* (pp 1053–1056).
4. Matei, Z., et al., (2012). Resilient distributed datasets: A fault-tolerant abstraction for in-memory cluster computing. *Proceedings of the 9th USENIX Symposium on Networked Systems Design and Implementation.*
5. Xin, L., et al., (2015). *Computational Trust Models and Machine Learning.* CRC Press.
6. James, S., (2004). *The Wisdom of Crowds, Why the Many are Smarter than the Few and How Collective Wisdom Shapes Business, Economies, Societies and Nations* (pp. 66–83). Anchor Books.
7. Paul, C. Z., et al., (2013). *Harness the Power of Big Data, The IBM Big Data Platform.* McGraw-Hill.
8. David, L., Ryan, K., Gary, K., & Alessandro, V., (2014). *The Parable of Google Flu: Traps in Big Data Analysis, Science* (Vol. 343, No. 6176, pp. 1203–1205).
9. Bhat, A. H., Patra, S., & Jena, D., (2013). "Machine learning approach for intrusion detection on cloud virtual machines." *International Journal of Application or Innovation in Engineering and Management (IJAIEM), 2*(6), 56–66. http://www.ijaiem.org/Volume2Issue6/IJAIEM-2013-06-09-029.pdf (accessed on 25 February 2020).
10. Chase, D. A., Thakar, P., Vahdat, A., & Doyle, R., (2001). "Managing energy and server resources in hosting centers." *18th ACM Symposium on Operating Systems Principles (SIGOPS), 35*(5), 103–116. http://dl.acm.org/citation.cfm?id=502045 (accessed on 25 February 2020).
11. Tierney, L., Rossini, A. J., & Na, L., (2009). Snow: A parallel computing framework for the R system. *Int. J. Parallel Prog., 37*, 78–90. doi: 10.1007/s10766-008-0077-2.
12. Hido, S., (2012). *Jubatus: Distributed Online Machine Learning Framework for Big Data, XLDB Asia.* Beijing. http://www.slideshare.net/JubatusOfficial/distributed-online-machine-learning-frameworkfor-big-data (accessed on 25 February 2020).
13. Ghoting, A., Kambadur, P., Pednault, E., & Kannan, R. (2011). *NIMBLE: A Toolkit for the Implementation of Parallel Data Mining and Machine Learning Algorithms on Map Reduce.* KDD 11.
14. Low, Y., Gonzalez, J., Kyrola, A., Bickson, D., Guestrin, C., & Hellerstein, J. M., (2012). Distributed GraphLab: A framework for machine learning and data mining in the cloud. *Proceedings of the VLDB Endowment* (Vol. 5, No. 8), Istanbul, Turkey.
15. Pednault, E., Yom-Tov, E., & Ghoting, A., (2012). IBM parallel machine learning toolbox. In: Bekkerman, R., Bilenko, M., & Langford, J., (eds.), *Scaling up Machine Learning.* Cambridge University Press.
16. Eckerson, W., (2012). *New Technologies for Big Data.* http://www.b-eyenetwork.com/blogs/eckerson/archives/2012/11/newtechnologie.php (accessed on 1 March 2020).

17. Shiravi, A., Shiravi, H., Tavallaee, M., Ghorbani, A., (2012). "Toward developing a systematic approach to generate benchmark datasets for intrusion detection." *Computers and Security, 31*(3), 357–374.

18. Sommer, R., & Paxson, V., (2010). "Outside the closed world: On using machine learning for network intrusion detection." In: *IEEE Symposium on Security and Privacy* (pp. 305–316).

19. Bhamare, D., Jain, R., Samaka, M., Vaszkun, G., & Erbad, A., (2015). "Multi-cloud distribution of virtual functions and dynamic service deployment: Open ADN perspective." In: *Cloud Engineering (IC2E)* (pp. 299–304). IEEE.

20. Mchugh, J., (2000). "Testing intrusion detection systems: A critique of the 1998 and 1999 DARPA intrusion detection system evaluations as performed by Lincoln Laboratory." *ACM Trans Inf. System Security.*

CHAPTER 5

Ontology-Based Information Retrieval and Matching in IoT Applications

M. LAWANYA SHRI,[1] E. GANGA DEVI,[2] BALAMURUGAN BALUSAMY,[3] and JYOTIR MOY CHATTERJEE[4]

[1]*School of Information Technology and Engineering, VIT, Vellore, India*

[2]*Loyola College, Chennai, Tamil Nadu, India*

[3]*School of Computer Science and Engineering, Galgotias University, Noida, Uttar Pradesh, India*

[4]*Department of IT, LBEF (APUTI), Kathmandu, Nepal*

ABSTRACT

Ontology is a knowledge depiction archetypal that provides the semantic services which are the most challenging task on the emergent cutting-edge technology such as IoT. The dynamic sensing capability with adapting the semantic structure to the sensed data gains precise attention from the research community. The semantic approach solves the issue of interoperability applied to the heterogeneous sensed data across different application domains. Generating ontologies from the heterogeneous sensed data sources based on real-time application needs an effective mechanism due to the rising demand for the public participation. We propose an effective technique for generating ontology by adopting the Fruit fly optimization algorithm. The proposed approach proves that the construction of ontology-based information increases the logistics of the system effectively with a low operating cost.

5.1   INTRODUCTION

The rapid growth of the internet of things provides a significant change with new areas like smart cities, smart homes, and so on. The sensitive information which is transmitted by the sensor devices from heterogeneous resources should be maintained and to focus on nurturing the inter-operability between different devices properly. Ontology-based on semantic structuring plays a vital role in IT applications that should strictly follow the uniformity of the data and semantics that are designated in an application-independent manner. Ontology deals with information processing of sensed data from heterogeneous resources for knowledge description and increases the logistic system effectively.

Ontology is the successor of metaphysics which is a branch of philosophy. This branch analyses existence, and its types or modes are frequently giving attention to the relations between particulars and universals, between essence and existence and between intrinsic and extrinsic properties. The goal of ontology is to discover fundamental categories or kinds into which objects naturally fall, by dividing the world "at its joints." It is a philosophical study of the nature of being, becoming, and reality. An ontology defines what entities exist and how entities are interrelated hierarchically. Here a database is selected and its table, column, and values in the cells are retrieved to develop ontology. Developing ontology involves owl file creation and ontology graph which shows the interrelationship between entities.

In the existing system algorithm for ontology development is given but for security/encryption, no algorithm is defined. However, integrating more IOT sensed data together through the internet leads to security issues and challenges. The proposed approach also focuses on the security issues by incorporating encryption and decryption algorithms with the proposed framework. For encrypting or hiding some information in the developed ontology, the encryption technique needs to be added. An information system is frequently using ontologies in the last decades. Natural language processing (NLP), Web technology, database integration, multi-agent systems, artificial intelligence (AI), etc. are the field where ontology is popular.

Semantic web, RDF, OWL are the terms which are a related and important part of the ontology. For data that are shared and reused transversely different communities and applications, a common framework is needed and Semantic web provides that framework [1]. Semantic Web defines a framework in which allows to share data across different platforms and application and is based on a metadata model knows as resource description framework (RDF) [2], which is a customary model for data exchange on the web. Large quantities of data should be stored in relational databases (RDB) so that it will be developed into the Semantic web. Developing RDF graph is a challenging task from a relational database that has been the goal of various researches.

RDF is a machine-readable language that describes subjects (resources), properties, and values. When it comes to mechanism on how to describe the properties, describe the relationship between properties and other resources, RDF has no mechanisms. Whereas, ontology explicitly gives a specification for the communal conceptualization of various domains and also describes various classes, properties, and relationships. Ontology usually defines the terms used in the RDF document with the use of Ontology description language called web ontology language (OWL) [3].

Here some rules have been represented and implemented to convert relational database into ontology. This approach will acquire instances from a relational database into ontologies. A prototype has been developed to create ontology from RDB. This will extract schema metadata of the relational database and OWL ontology is created [4]. The data of RDF is stored in an OWL document. Encryption mechanism is also added to encrypt data in the OWL file and in Ontology graph. Day by day, there is increase in the internet database or offline database and information is stored repeatedly and used again and again. So to provide a common framework to share data and use it in different domains ontology comes in the scene [5]. Ontology provides common vocabulary for information sharing, for example, medical websites having information about medicines, if these websites share and publish common words/vocabulary of medicines then computer agents can extract and aggregate information from these websites for user queries.

The proposed system has an ontology development algorithm as well as an encryption/decryption algorithm. By using encryption user can develop ontology and encrypt whatever information he wants. AES encryption algorithm is used in the proposed system. A key will be used for encryption and decryption which will be created by admin (Figure 5.1).

Here the output is different because three websites are having different names for same medicines but using ontology software agent will identify what user wants and it will collect information from all websites using common vocabulary. So in the paper methods/algorithm for ontology is given but for how it will be implemented, is not described. There are several methods described for converting relational database into ontology (RDF) but none of the methods has encryption techniques to secure data. Here the method to implement paper as well as providing encryption mechanisms for securing private data is implemented.

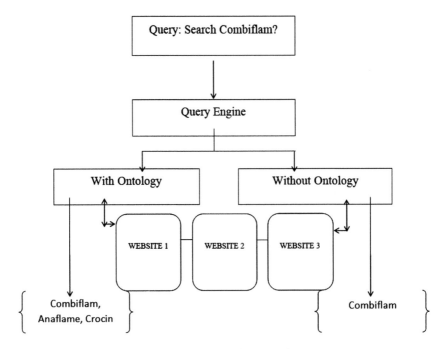

FIGURE 5.1   Difference in output with and without ontology.

Ontology matching technique in this approach uses fruit fly optimization algorithm which is a global optimization algorithm inspired by the foraging behavior of fruit flies [6]. Fruit fly optimization approach is used in the ontology-based IOT application due to the issues like poor automation, slow convergence, and excess of manual intervention to convert from relation database to OWL and for interoperability issues [7]. FOA method chosen because there is less number of parameters involved in the algorithm that makes the algorithm to implement and understand easily. The application of FOA approach proved that it is better compared to other optimization algorithm. The exploitation ability of fruit fly optimization makes the algorithm more standard that can be achieved using local searching process. Fruit fly optimization technique is a nature inspired algorithm which uses two searching strategy [8]. First, it uses smell based approach to identify the location of rotten fruits and fly towards the direction, and then secondly it used a vision-based approach for selecting the best fruit among the other fruits [9]. The performance capability with better convergence and stability nature was the reason to choose the FOA algorithm for ontology in IoT applications [10].

5.2 RELATED WORK

In this proposed technique, we make use of the ontology which provides the latter classification that is based on the functionalities. Our objective is not only on the specific application, but our desire is to model the whole IoT domain. And hence, we limit our discussion to generic and domain-specific ontologies.

These ontologies are extensively used in different application areas such as building management systems (BMS) [11], indoor navigation [12], and smart-homes [13]. Ye et al. [14] discussed about a categorization of ontologies which are based on expressiveness as light-weight or heavy-weight or generality as generic, domain-specific, or application-specific.

Kelly et al. [15] introduce an efficient smart home framework in which the context-aware reasoning to monitor systematic domestic environments and to aggregate data. In paper [16], DB2OWL is described by the authors by converting tables to concepts and columns to predicates and the document is saved in R2O. Here no mechanism is given for mapping instances. In paper [17], a D2R approach used a XML-based language where the mapping between ontologies implemented in RDF Schema and relational database model is described. But how to show the ontology graph is not described.

For constructing flexible data, a semantic model the combination of RDF/OWL has been used [18]. But in the proposed approach RDF graph is generated from directly and automatically from relational database. Nasser et al. [19] presented an efficient approach is used to develop ontology from existing database for database integration. Database from many sources can be integrated by using ontology is shown in this research work. Irina et al. proposed a method where ontology from owl file is converted into RDB.

In this research, work [20] a utility is developed called "qualeg" which selects ontology and converts it into SQL statement and on executing that SQL statement ontology is created. In this paper [21] describes how a common vocabulary can be developed and used semantically. By using common names for food, interested parties will have better integration and collaboration. The ontology developed is compatible with SPARQL and SPARQL queries can be executed to retrieve information from food ontology.

In this, paper [22] applying rules have been defined to map relational database to ontology. Approach to take database data is also implemented using an application. Automatic conversion of constructs of relational data model to OWL ontology is focused here. In this paper [23] the entities that are ranked using semantic structure is presented effectively. The semantic resources that are aggregated are sorted based on the parameter value relevant to the overall ranking with the query length [24].

The semantic related viewpoint is due to the issues such as automation, data analytics, and interoperability in IOT applications [25]. The limitation of interoperability across ontologies in IOT application is due to the heterogenetic nature for knowledge representation of semantic structure within the same domain [26]. In this paper [27], the author discussed various optimization algorithms used for ontology matching problems and drawbacks of evolutionary algorithms.

This paper [28] presented a framework by using ordered weighted average operation for ontology matching. The position with the same values in the ranking list is obtained to determine the matching elements in the same domain [29]. In this paper [30], Minkowski distance is obtained to measure the likeness of two identical matrices by determining the weight of similar matching elements based on ranking.

Abdullah et al. [31] proposed symbiotic Organism search optimization to perform scheduling in a cloud environment to minimize the make span, response time and the imbalance degree. Liang Hong et al. [32] represented a cloud model based FOA to improve the performance in a better way.

Ling et al. [33] designed a novel based FOA for proving the multidimensional knapsack problem (MKP). Global vision-based, Smell-based, and local vision-based searching techniques are implemented to fulfill the evolutionary search algorithm. FOA uses a binary string representation to solve MKP.

Hong et al. [34] proposed a new hybrid annual power load forecasting model which is joined with the fruit fly algorithm (FOA) and generalized regression neural network (GRNN). To improve the forecasting accuracy of GRNN in power load annual forecasting, a technique called FOA is used to choose spread parameter value.

Peng et al. [35] verified an improvement in the fruit fly optimization algorithm to search the result for lot-streaming flow shop scheduling problem. This method made use of neighborhood search and global cooperation search process to locate the jobs splitting and order of sub-lots immediately.

Zheng et al. [36] presented a FOA to solve semiconductor final testing scheduling. Kennedy et al. [37] used the Particle Swarm Optimization algorithm and explained their application in various fields.

This paper [38] handles the situation by imparting M3 ontology-an aspect of ontology introduced as a supplement to W3C's SSN ontology which handles the description of sensors, observations, phenomena, and domains which permits for reasoning on sensor data using rules to infer contextual information.

Because, the sensors are attached with the mobile device, the problem can raise for dynamicity and the sensor discovery. This kind of problems can also be handled here. These kinds of issues were taken into OntoSensor [39]. The ontology concepts expand towards SensorML 9, ISO-19115 10, and SUMO [40]. These terminologies accept for identification of sensor distribution, its conduct, relationships, quality, and meta-data related to characteristics of sensor, its behavior, and reliability. OntoSensor focuses to balance interoperability and inferences of ontology which needs the characteristics of physical sensing that is to be incorporated in the explanation of ontology.

The ontology is too weight, which has the usage complexity and hence it is inadequate to define the observation of sensors. This issue was answered by MyOntoSens [41] by issuing a generic and exhaustive ontology which explains sensor observations and the potentials to give the collected information.

MyOntoSens is discovered for the area of wireless sensor networks (WSN) and receives various concepts and relationships from the existing ontologies which add OntoSensor [42], SSN [43], and QUDT [44]. The mentioned ontologies can be applied for the domain of IoT also. MyOntoSens issues the methodology to balance the sensor discovery and sensor registration to work in the domain. Hirmer et al. [45] explained ontology to help hand for dynamic registration and bindings of new sensors to a platform.

They gain the concepts of sensors and their related properties from SensorML which showed additional information of Adapter clubbed with every sensor. The existing concepts support sensor discovery and the proposed method supports introduced concepts and the additional information about sensor data.

The updated researches on the Internet of Things are maintained in three areas. They are: things, networks, and intelligence. From these areas, the priority is given to the frameworks of IoT, which focuses on the visibility of improvement of things as identification and investigation. The famous architecture is EPCGlobal [46] explained by Auto-ID center of MIT and UID [47] who used Ubiquitous ID Center. On the other hand, the proposed method includes WSN, privacy protection, and security. Various studies are committed to increase the efficiency of wireless sensor network capacity, scalability, reliability, and robustness [48] such as 6LoWPAN [49] which are combined with the IP protocol and IEEE 802.15.4. It is a lightweight security architecture of networks and smart identification. In the area of intelligence, many researchers identified to make use of distributed AI, rich intelligent application (service) to facilitate the problems of coordinating interaction between tasks of IoT [50].

Also, the most emerging phase of IoT is to enhance a suitable architecture, where the functions, data process, and message transfer models can be designed to model and control the transactions. In the existing system, the research based on the semantic web was made to join the AI and knowledge engineering to demonstrate and to process the data and knowledge. The meaningful technology with all the descriptions by machines provides a method to explain the heterogeneous objects, information sharing, and integration issues. The semantic web technologies are ontology, semantic annotation, bond information, and semantic web services [51], which are used as an important solution to observe the sharing of semantic information between IoT entities. Of all these, the various studies are Task Computing based middleware [52], Smart Semantic middleware [53], and Semantic Device bus [54].

In 2009, Scioscia, and Ruta discussed to use the technologies of the semantic web with IoT and developed the semantic web of things (SWoT) [55]. The smart way is to apply the theory of ontology to make IoT intelligent. The University of Munich and DOCOMO Euro-Labs developed the Perci framework to implement basic mechanisms for seamless interaction between mobile services and tagged objects [56].

5.3   PROPOSED SYSTEM

"Ontology-based access to information system" in this an application is developed which will take the database as an input and give OWL file and RDF graph (Ontology Graph) as output.

For converting the database into algorithms are given. Tables are converted into classes, a foreign key is used for object properties in owl, and the primary key is used for data properties. AES algorithm is implemented for encrypting data in relational database tables.

5.4   BASIC ARCHITECTURE

5.4.1   OWL GENERATION

OWL file generation is not a single step or process it involves multiple steps one after another. First, according to rules table names are mapped to class names considering class and subclass. Second, data types of columns are mapped to data properties. Third, according to foreign key and primary key object property is mapped. Fourth, instances are generated (Figure 5.2).

5.4.2   ONTOLOGY GRAPH GENERATION

After saving, the owl script under xml tags in a. Owl file protégé editor is used. Protégé editor is ontology editor [57], here I am using to open. owl file and view the Ontology graph. The graph opened in the ontology editor will show the entities and their relationship, their hierarchy, number of successors, predecessors, etc. The graph can be saved in an image file (jpeg, bmp, etc). An encryption key will be used by admin, which will be used for decryption also since here symmetric encryption is used. The algorithm used is AES. Encrypted values will appear in the Ontology owl file and graph but in encrypted form [58] (Figure 5.3).

5.5   FRUIT FLY OPTIMIZATION ALGORITHM

Fruit fly optimization algorithm is an intelligent search algorithm that implements the food search procedure which is based on the drosophila's fruit fly behavior as shown in Figure 5.4. The osphresis organs present in fruit fly which situate several forms of scents floating and it can smell the gatherings of food sources even which is very far away from it. Fruit fly assembles and sends information from its adjacent fruit, compares, and searches for the best vision and fitness by taste. If the taste is not good, it gets rid off and moves to another location until it is satisfied with its best optimal solution. For the process of searching food, the two steps are used.

- It uses its osphresis organ which smells the food and moves to that location; and
- Fruit flies follow a sensitive vision to find its target food and its congregate location.

Step 1: Initialize the variables; Assign the population size, starting location of the fruit fly, and a maximum number of generations.

$$X_K, Y_K, Dis_K, S_K, Smell_K, \text{X-axis, Y-axis, Random value}$$

Step 2: Using the Olfactory organ behavior, assign each fruit fly with the direction and distance to search the food randomly

$$X_k = X - axis + Random\ Value$$
$$Y_k = Y - axis + Random\ Value$$

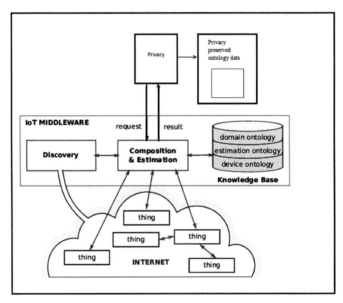

FIGURE 5.2   Ontology for IoT applications.

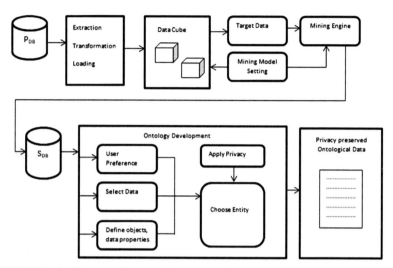

FIGURE 5.3 Architecture diagram of secure ontology-based data access.

Step 3: Assess the concentration of smell fitness value of each food location to estimate the distance of food source.

$$Dis_k = \sqrt{X_{k^2} + Y_{k^2}}$$

$$SmellConcentration(S_k) = 1 / Dis_k$$

Step 4: Find smell concentration of each fruit fly by replacing smell concentration.

$$Smell_k = Smell\_Function(S_k)$$

Step 5: Recognize the best smell concentration which has minimum value fruit fly.

$$[Best\_Smell, Best\_index] = \min(Smell_k)$$

Step 6: Use the vision-based search to find the direction of location with the maximum smell concentration on X and Y coordinates.

$$Smell\_Best = Best\_Smell$$
$$X\_axis = X(Best\_index)$$
$$Y\_axis = Y(Best\_index)$$

The vision and search based global optimization technique are very effective due to the double-blinded search strategy adopted in the fruit fly optimization algorithm. It is suitable for this proposed work for information processing and integration of ontologies taken from IOT sensed data. Figure 5.4 illustrates the flow diagram of the fruitfly optimization algorithm.

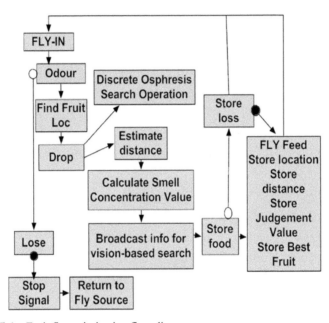

FIGURE 5.4 Fruit fly optimization flow diagram.

Algorithm 1: The Basic Procedure for FOA

Input: Population size, initial Fruit fly location, the maximum number of iterations.

Output: Best solution.

1. Do.
2. For (All the fruit in various positions).
3. Initialization.
4. Allocate route and distance to travel.
5. Estimate the smell concentration fitment rate.
6. Substitute the estimated smell value into the fitness method.
7. Find the best one among the calculated smell by maximum smell.
8. Vision search is used to identify the best smell location.
9. End For.
10. While stopping condition is not exceeded.

Proposed Work

Ontologies are currently used in the IoT application for knowledge representation and for structuring the sensed information of a specific domain. The main objective of this chapter is for designing an information processing model for IoT applications based on the semantic web and for integrating ontologies for enriching the information gathered from the sensors. Semantic technology plays a vital role in combining the entities, gathering, and monitoring the data.

Algorithm 1: The Basic Procedure for FOA

Input: Population size, initial Fruit fly location, the maximum number of iterations.

Output: Best solution.

11. Do.

12. For (All the fruit in various positions).

13. Initialization.

14. Allocate route and distance to travel.

15. Estimate the matching document using the concentration fitment rate.

16. Substitute the estimated matching ration of the document into the fitness method.

17. Find the best matching document for information retrieval by maximum ratio.

18. Vision search is used to identify the best ontologies (smell location).

19. End For.

20. While stopping condition is not exceeded.

Ontologies for IoT applications are used for annotating the data which is acquired from sensor entities for improving the efficiency of retrieval, processing, and integrating data. The proposed approach uses a fruitfly optimization technique to generate rank measures related to the similarity measures. The ranking is used to determine the ontologies having similar measures for improving the ontology matching and for integrating the ontologies. Each fly from the population in a particular domain first uses smell based approach to select the similarity and a vision-based approach is used to select most matching information.

5.6   CONCLUSION AND FUTURE WORK

"Ontology-based access to information" is implemented successfully and is able to give output. The key aspects such as OWL file, Ontology graph generation took lot of time and resource for converting their algorithm into the application. The overall conclusion one can draw using ontology one can very easily and efficiently show the entities and their relationship and that the application is able to do. Fruit fly optimization is used for Ontology matching and the results prove that it is more efficient and cost-effective. The feature or module for graph generation still needs an extra tool called Protégé that can be included as a module to show the graph from the generated OWL script. A module for running SPARQL (query for ontology) as a GUI can be added to the present application. SPARQL is RDF query language, Using SPARQL one can retrieve and manipulate data stored in RDF format. It is the key technology of semantic web. In SPARQL, queries have triple patterns, conjunction, disjunction, and optional patterns. The ontology editor used called protégé has option to run SPARQL, since protégé is developed in java so the API's for SPARQL can be used for developing the query module.

KEYWORDS

- fruitfly optimization
- ontology
- ontology graph generation
- osphresis organ
- OWL generation
- web ontology language

REFERENCES

1. Grigoris, A., & Frank, V. H., (2004). *A Semantic Web Primer*. The MIT Press, Cambridge Massachusetts, London, England.
2. Klyne, G. (2004). Resource description framework (RDF): Concepts and abstract syntax. *http://www.w3.org/TR/2004/REC-rdf-concepts-20040210/*.

3. McGuinness, D. L., & Van Harmelen, F. (2004). OWL web ontology language overview. *W3C Recommendation, 10*(10), 2004.
4. Gruber, T. R., (1993). "A translation approach to portable ontology specifications." *Knowledge Acquisition, 2*, 199–220. doi: 10.1006/knac.1993.1008.
5. Hepp, M. (2007). Possible ontologies: How reality constrains the development of relevant ontologies. *IEEE Internet Computing, 11*(1), 90–96.
6. Pan, W. T., (2011). A new evolutionary computation approach: Fruit fly optimization algorithm. In: *Proceedings of the Conference on Digital Technology and Innovation Management*.
7. Lin, S. M., (2013). Analysis of service satisfaction in web auction logistics service using a combination of fruit fly optimization algorithm and general regression neural network. *Neural Computing and Applications, 22*(3/4), 783–791.
8. Wang, L., Zheng, X. L., & Wang, S. Y., (2013). A novel binary fruit fly optimization algorithm for solving the multidimensional knapsack problem. *Knowledge-Based Systems, 48*, 17–23.
9. Li, H. Z., Guo, S., Li, C. J., & Sun, J. Q., (2013). A hybrid annual power load forecasting model based on generalized regression neural network with fruit fly optimization algorithm. *Knowledge-Based Systems, 37*, 378–387.
10. Zhang, P., & Wang, L., (2014). Grouped fruit-fly optimization algorithm for the no-wait lot streaming flow shop scheduling. In: *International Conference on Intelligent Computing* (pp. 664–674). Springer International Publishing.
11. Ye, J., Coyle, L., Dobson, S., & Nixon, P., (2007). *Ontology-Based Models in Pervasive Computing Systems, the Knowledge Engineering Review, 22*(04), 315–347.
12. Sarma, A. C., & Girao, J., (2009). "Identities in the future internet of things." *Wireless Personal Communications, 49*(3), 353–363.
13. Atzori, L., Iera, A., & Morabito, G., (2010). *The Internet of Things: A Survey, Computer Networks, 54*(15), 2787–2805.
14. Song, Z., Cardenas, A. A., & Masuoka, R., (2010). "Semantic middleware for the internet of things." In: *Proceedings of the 2ⁿᵈ International Internet of Things Conference (IoT '10)* (pp. 1–8). Tokyo, Japan.
15. Katasonov, A., Kaykova, O., Khriyenko, O., Nikitin, S., & Terziyan, V., (2008). "Smart semantic middleware for the internet of things." In: *Proceedings of the 5ᵗʰ International Conference on Informatics in Control, Automation and Robotics (ICINCO '08)* (pp. 169–178). Funchal, Portugal.
16. Vasseur, B., Van De Vlag, D., Stein, A., Jeansoulin, R., & Dilo, A., (2004). "Spatio-temporal ontology for defining the quality of an application." In: *Proceedings of the International Symposium on Spatial Data Quality* (pp. 67–81), Bruck an der Leitha, Austria.
17. El-Geresy, B. A., Abdelmot, A. I., & Jones, C. B., (2002). "Spatiotemporal geographic information systems: A causal perspective." In: *Advances in Databases and Information Systems* (Vol. 2435, pp. 191–203), Springer, Berlin, Germany.
18. Martin, D., Burstein, M., Hobbs, J., et al., (2004). "*Owl-s: Semantic Markup for Web Services.*"http://www.ai.sri.com/daml/services/owl-s/1.2(accessedon25February2020).
19. Kelly, S. D. T., Suryadevara, N. K., & Mukhopadhyay, S. C., (2013). Towards the implementation of IOT for environmental condition monitoring in homes. *IEEE Sens. J., 13*(10), 3846–3853.

20. Nadine, C., Raji, G., & Kokou, Y., (2007). "DB2OWL: A tool for automatic database-to-ontology mapping." In: *Proceedings of 15ᵗʰ Italian Symposium on Advanced Database System (SEBD 2007)* (pp. 491–494).
21. Christian, B., (2003). "D2R MAP: A database to RDF mapping language." In: *Proceedings of the 12ᵗʰ International World Wide Web Conference (WWW2003)*.
22. Heru, A. S., Su-Cheng, H., & Ziyad, A. M. T., (2011). "Ontology extraction from relational database: Concept hierarchy as background knowledge." *Knowledge-Based Systems, 24*(3), 457–464.
23. Kārlis, Č., & Guntars, B., (2011). "RDB2OWL: A RDB-to-RDF/OWL mapping specification language." In: *Proceedings of the 2011 conference on Databases and Information Systems VI: Selected Papers from the Ninth International Baltic Conference (DB&IS 2010)* (pp. 139–152). IOS Press.
24. Munir, K., Odeh, M., & McClatchey, R., (2012). "Ontology-driven relational query formulation using the semantic and assertional capabilities of OWL-DL." *Knowledge-Based Systems, 35*, 144–159.
25. Nasser, A., Hussein, Z., & François, S., (2009). *"Generating OWL Ontology for Database Integration"* Third International Conference on Advances in Semantic Processing. Leicester, UK.
26. Irina, A., Nahum, K., & Ahto, K., (2007). "Storing OWL ontology's in SQL relational databases." *International Journal of Computer, Electrical, Automation, Control, and Information Technology Engineering, 1*(5), 1261.
27. Maxim, K., & Dmitry, Z., (2009). "Food product ontology: Initial implementation of a vocabulary for describing food products." *Proceeding of the 14ᵗʰ conference of Fruct Association*. Saint-Petersburg, Russia.
28. Zdenka, T., (2010). "Relational database as a source of ontology creation." *Proceedings of the International Multi Conference on Computer Science and Information Technology* (pp. 135–139). Ostrava Czech Republic.
29. Graves, A., Adali, S., & Hendler, J., (2008). A method to rank nodes in ARDF graph. In: Bizer, C., & Joshi, A., (eds.), *International Semantic Web Conference (Posters and Demos), in: CEUR Workshop Proceedings* (Vol. 401), CEUR-WS.org.
30. Barnaghi, P., Wang, W., Henson, C., & Taylor, K., (2012). "Semantics for the internet of things: early progress and back to the future." *International Journal on Semantic Web and Information Systems, 8*(1), 1–21.
31. Abdullahi, M., & Ngadi, M. A., (2016). Hybrid symbiotic organisms search optimization algorithm for scheduling of tasks on cloud computing environment. *PLOS One, 11*(6), e0158229.
32. Črepinšek, M., Liu, S. H., & Mernik, L., (2012). A note on teaching-learning-based optimization algorithm. *Information Sciences, 212*, 79–93.
33. De, S., Barnaghi, P., Bauer, M., & Meissner, S., (2011). "Service modeling for the Internet of Things." In: *Proceedings of the Federated Conference on Computer Science and Information Systems (FedCSIS '11)* (pp. 949–955). IEEE.
34. Nadine, C., Raji, G., & Kokou, Y., (2007). "DB2OWL: A tool for automatic database-to-ontology mapping." In: *Proceedings of 15ᵗʰ Italian Symposium on Advanced Database System (SEBD 2007)* (pp. 491–494).
35. Pan, W. T., (2012). A new fruit fly optimization algorithm: Taking the financial distress model as an example. *Knowledge-Based Systems, 26*, 69–74.

36. Han, J., Wang, P., & Yang, X., (2012). Tuning of PID controller based on fruit fly optimization algorithm. In: *2012 IEEE International Conference on Mechatronics and Automation* (pp. 409–413). IEEE.

37. Choubey, N. S., (2014). Fruit fly optimization algorithm for travelling salesperson problem. *International Journal of Computer Applications (0975–8887), 107*(18), 22–27.

38. Li, J. Q., Pan, Q. K., Mao, K., & Suganthan, P. N., (2014). Solving the steelmaking casting problem using an effective fruit fly optimization algorithm. *Knowledge-Based Systems, 72,* 28–36.

39. Mousavi, S. M., Alikar, N., Niaki, S. T., & Bahreininejad, A., (2015). Optimizing a location allocation-inventory problem in a two-echelon supply chain network: A modified fruit fly optimization algorithm. *Computers and Industrial Engineering, 87,* 543–60.

40. Zhang, Y., Cui, G., Wang, Y., Guo, X., & Zhao, S., (2015). An optimization algorithm for service composition based on an improved FOA. *Tsinghua Science and Technology, 20*(1), 90–99.

41. He, X., & Baker, M., (2010). xhRank: Ranking entities on the semantic web. In: Polleres, A., Chen, H., (eds.), *ISWC Posters & Demos, in: CEUR Workshop Proceedings* (Vol. 658). CEUR-WS.org.

42. Atzori, L., Antonio, I., & Giacomo, M., (2010). "The internet of things: A survey." *Computer Networks, 54*(15), 2787–2805.

43. Rangel, C., Aguilar, J., Cerrada, M., & Altamiranda, J., (2015). "An approach for the emerging ontology alignment based on the bees colonies." *Submitted to International Joint Conference on Artificial Intelligence (IJCAI).*

44. Xue, X., & Wang, Y., (2015). Optimizing ontology alignments through a memetic algorithm using both match F-measure and unanimous improvement ratio. *Artif. Intell., 223,* 65–81.

45. Ehrig, M., (2007). *Ontology Alignment: Bridging the Semantic Gap, Semantic Web and Beyond Computing for Human Experience* (Vol. 4). Springer, Heidelberg.

46. Yager, R. R., (1988). On ordered weighted averaging aggregation operators in multicriteria decision making. *IEEE Trans. Syst. Man Cybern., 18*(1), 183–190.

47. Yager, R. R., (1992). Applications and extensions of OWA aggregations. *Int. J. Man Mach. Stud., 37*(1), 103–122.

48. Yager, R. R., (1993). Families of OWA operators. *Fuzzy Sets Syst., 59*(2), 125–148.

49. Pfisterer, D., Romer, K., Bimschas, D., et al., (2011). "SPITFIRE: Toward a semantic web of things," *IEEE Communications Magazine, 49*(11), 40–48.

50. Noy, N. F., (2001). *"Ontology Development 101: A Guide to Creating Your First Ontology: Knowledge Systems Laboratory."* Tech. Rep. KSL-01-05, SMI-2001-0880, Stanford University. Stanford Knowledge Systems Laboratory, Stanford Medical Informatics, Stanford, Calif, USA.

51. Dibowski, H., & Kabitzsch, K., (2011). Ontology-based device descriptions and device repository for building automation devices. *EURASIP J. Embedded Syst.*

52. Gyrard, A., Bonnet, C., Boudaoud, K., & Serrano, M., (2016). LOV4IoT: A second life for ontology-based domain knowledge to build semantic web of things applications. In: *Future Internet of Things and Cloud (FiCloud), 2016 IEEE 4ᵗʰ International Conference* (pp. 254–261). IEEE.

53. Yan, B., Hu, Y., Kuczenski, B., Janowicz, K., Ballatore, A., Krisnadhi, A. A., Ju, Y., Hitzler, P., Suh, S., & Ingwersen, W., (2015). *An Ontology for Specifying Spatiotemporal Scopes in Life Cycle Assessment* (pp. 25–30). In: Diversity++@ ISWC.

54. Niitsuma, M., Yokoi, K., & Hashimoto, H., (2009). Describing human-object interaction in intelligent space. In: *Human system interactions, 2009. HSI'09* (pp. 395–399). 2nd Conference on, IEEE.

55. Zhang, D., Wang, L., Xiong, H., & Guo, B., (2014). 4W1H in mobile crowd sensing. *IEEE Communications Magazine, 52*(8), 42–48.

56. Wang, W., De, S., Cassar, G., & Moessner, K., (2013). Knowledge representation in the internet of things: Semantic modeling and its applications. *Automatika, 54*(4), 388–400.

CHAPTER 6

Parts-of-Speech Tagging in NLP: Utility, Types, and Some Popular POS Taggers

SOUMITRA GHOSH[1] and BROJO KISHORE MISHRA[2]

[1]Department of Computer Science and Engineering, Indian Institute of Technology Patna, India

[2]Department of Computer Science and Engineering, GIET University, Gunupur, Odisha, India

ABSTRACT

Majority of the basic models in the area of Natural Language Processing (NLP) are based on Bag of Words. Primary limitation of Bag of Words based models is their inability to capture the syntactic relations between words. Part-of-Speech (POS) are uselful in improving on this Bag of Words technique. Part-of-Speech (PoS) tagging is involved with the process of assigning one of the parts of speech (include nouns, verb, adverbs, adjectives, pronouns, conjunction and their sub-categories) to the given word. POS tagging is the process of marking up a word in a corpus to a corresponding part of a speech tag, based on its context and definition. POS Tags are useful for building parse trees, which are used in building Named Entity Recognitions (NERs) and extracting relations between words. POS Tagging also has major application in building lemmatizers which are used to reduce a word to its root form. To understand the meaning of any sentence or to extract relationships and build a knowledge graph, POS Tagging is a very important step. This chapter gives a vivid introduction to the POS tagging problem, its various applications and types with special emphasis on markov model based POS tagging and finally some python based implementtaion of some popular POS taggers.

6.1 INTRODUCTION

Parts-of-Speech (otherwise called POS, word classes, or syntactic classifications) are helpful on account of the substantial measure of information they give about a word and its neighbors. To know if a word is a noun or a verb discloses to us a lot about likely adjacent words (nouns are placed before by determiners and adjectives, verbs by nouns) and about the syntactic structure around the word (nouns are usually part of noun phrases), which makes parts-of-speech labeling an imperative segment of syntactic parsing. Parts-of-speech are valuable features for finding named entities like individuals or organizations in texts and other data extraction assignments. Parts-of-speech impact the conceivable morphological attaches thus can impact stemming for data retrieval, and can help in summarization for enhancing the choice of nouns or other vital words from an archive. A word's parts-of-speech are imperative for producing pronunciations in speech synthesis and recognition. The word 'compact,' for instance, is articulated COMpact when it is a noun and comPACT when it is an adjective [1].

From our early childhood, we have been made acquainted with recognizing parts of speech labels. For instance, reading a sentence and having the capability to recognize what words go about as nouns, pronouns, verbs, adverbs, etc. All these are alluded to as the parts-of-speech tags.

6.2 BASICS OF POS TAGGING

6.2.1 PARTS-OF-SPEECH TAGGING

The part of speech explains how a word is used in a sentence. There are eight main parts of speech-nouns, pronouns, adjectives, verbs, adverbs, prepositions, conjunctions, and interjections. Labeling a word to one of the parts of speech tags is known as Parts of Speech tagging, also referred to as POS tagging [2].

Example: Word: *Fish*, POS Tag: *Noun*, Word: *Beautiful*, POS Tag: *Adjective*, Word: *Jump*, POS Tag: *Verb*, Word: *And*, POS Tag: *Conjunction*, etc.

It is to be noted that a word can have multiple tag values associated with it based on the context it is appearing. For example, the paddle can be a noun as well as a verb. Also, attempt, bear, face, dust, act, etc are some words with both nouns as well as verb tags.

Distinguishing parts-of-speech labels is substantially more confounded than basically mapping words to their parts-of-speech labels. This is on the grounds that POS labeling is not something that is non-exclusive. It is very feasible for a particular word to have an alternate parts-of-speech tag in various sentences dependent on various contexts. That is the reason it is difficult to have a conventional mapping for POS labels. As should be obvious, it is beyond the realm of imagination to physically discover diverse parts-of-speech labels for a given corpus. New kinds of contexts and new words keep coming up in dictionaries in different languages, and manual POS labeling is not versatile in itself. That is the reason we depend on machine-based POS labeling.

6.2.2 PARTS-OF-SPEECH TAGGER

Parts of Speech tagger or POS tagger is a program that carries out POS Tagging. Taggers utilize a various types of data: lexicons, dictionaries, rules, etc. A word may have a place with more than one class in a dictionary. For instance, 'skate' can be a noun as well as a verb based on its usage. Taggers utilize probabilistic data to comprehend this uncertainty.

There are fundamentally two sorts of taggers: rule-based and stochastic. Rule-based taggers use manually written standards to recognize the ambiguity in tags [3]. Stochastic taggers are either hidden Markov model (HMM) based [4], picking the tag sequence which increases the result of word probability and tag sequence likelihood, or cue-based, utilizing decision trees (DTs) or maximum entropy models to join probabilistic features.

Preferably, a regular tagger ought to be powerful, effective, precise, tunable, and reusable. In general, taggers either right away recognize the tag for the given word or make the best speculation dependent on the accessible data. As the normal language is complex and unpredictable, it is often troublesome for the taggers to settle on exact choices about labels. So incidental blunders in tagging are not taken as a noteworthy detour to look into.

6.2.3 TAGSET

Tagset is the collection of tags from which the tagger finds appropriate tags and attaches to the pertinent word. Each tagger will be given a standard tagset. The tagset might be a small one, for example, N (noun), V (verb), ADJ (adjective), ADV (adverb), PREP (preposition), CONJ (conjunction) or fine-grained, for example, NNOM (noun-nominative), NSOC (noun-sociative), VFIN (verb finite), VNFIN (verb nonfinite), etc. A large portion of the taggers utilizes only fine-grained tagset.

6.2.4 ARCHITECTURE OF POS TAGGER

1. **Tokenization:** The given content is partitioned into tokens with the goal that they can be utilized for further investigation. The tokens might be words, punctuations, and utterance limits [2].
2. **Ambiguity Look-Up:** This is to utilize lexicons and a predictor (guessor) for obscure words. While the dictionary gives a rundown of the list of words and their conceivable parts-of-speech tags, predictors examine obscure tokens. A lexical analyzer is comprised of a compiler or interpreter, lexicon, and predictor.
3. **Ambiguity Resolution:** Disambiguation is the other name of Ambiguity Resolution [2]. Disambiguation depends on information about word, for example, the likelihood of the word. Disambiguation is also based on the context in which a word happened to occur. For instance, the model may incline toward examining nouns over verbs if the former word is a preposition or article. Disambiguation is quite a troublesome issue in parts-of-speech tagging.

6.3 APPLICATIONS OF POS TAGGING

Parts-of-speech tagging in itself may not be the answer for a specific NLP issue. It is anyway something that is done as a pre-essential to rearrange a variety of issues. Some of its important applications in various NLP tasks are:

6.3.1 TEXT TO SPEECH CONVERSION

POS tagging finds huge applications in text-to-speech systems [5]. Let us consider the following sentence:

'Please resume reading my resume.'

Let's have a look at the various POS tags generated for the above sentence using NLTK package.

- **Python code snippet:**

  ```
  >>> text = word_tokenize('Please resume reading my resume.')
  >>> text
  ['Please,' 'resume,' 'reading,' 'my,' 'resume,'.']
  >>>nltk.pos_tag(text)
  ```

Output:

```
[('Please,' 'NNP'), ('resume,' 'VB'), ('reading,' 'VBG'), ('my,'
'PRP$'), ('resume,' 'NN'), (.,'.')]
```

The word 'resume' is being utilized twice in this sentence and has two distinct implications here. First resume is a verb signifying 'continue,' while second resume is a noun signifying a document (that is, they are not homophones). Along these lines, we have to realize which word is being utilized so as to articulate the content effectively. As we can see that a word may have different POS labels, it is pretty obvious for a text-to-speech converter to come up with a different set of sound for such words.

6.3.2 *WORD-SENSE DISAMBIGUATION (WSD)*

Words frequently happen in various contexts as various parts of speech. For instance, consider the following sentence:

'Some equations are easy to expand. Rails expand in summer.'

- **Python Code Snippet:**

  ```
  >>> text = word_tokenize('Some equations are easy to expand.
  Rails expand in summer.')
  >>> text
  ['Some,' 'equations,' 'are,' 'easy,' 'to,' 'expand,'.,'' 'Rails,' 'expand,'
  'in,' 'summer,'.']
  >>>nltk.pos_tag(text)
  ```

Output:

```
[('Some,' 'DT'), ('equations,' 'NNS'), ('are,' 'VBP'), ('easy,' 'JJ'),
('to,' 'TO'), ('expand,' 'VB'), (.,'.'), ('Rails,' 'NNPS'), ('expand,'
'NN'), ('in,' 'IN'), ('summer,' 'NN'), (.,'.')]
```

The word 'expand' in the above sentences has totally unique references, yet more essentially one is a noun and the other is a verb. Simple word-sense disambiguation (WSD) is conceivable in the event that you can label words with their POS labels.

WSD [6] is recognizing which context of a word (that is in which sense) is utilized in a sentence, when the word has various implications.

These are only two of the various applications where we would require parts-of-speech tagging [7]. Other applications where POS tagging is extensively used are question answering, speech recognition, machine translation, named entity recognition (NER), etc.

6.4   TYPES OF POS TAGGERS

Based on the techniques of POS tagging, POS taggers can be classified into two distinctive groups: Rule-Based POS Taggers and Stochastic POS Taggers.

6.4.1   RULE-BASED POS TAGGER

The automatic part of speech tagging is an area of NLP where statistical techniques have been more useful than rule-based methods. Typical rule-based methodologies utilize contextual data to allot tags to obscure or ambiguous words. Disambiguation is performed by analyzing the linguistic features of the word, its previous word, the next word, etc. For instance, on the off chance that the previous word is an article, the word being referred to must be a noun. This information is coded as rules or guidelines.

A rule might be like: *On the off chance that an uncertain/obscure word X is preceded by a determiner and succeeded by a noun, label it as an adjective.*

It is tiresome to formulate the set of rules for a tagger as it would require a lot of time and human efforts. So we require some programmed methods for doing this. The Brill's tagger is a rule-based tagger that helps to find the set of rules for tagging according to the type of data and limiting POS tagging mistakes. The most vital point to note here about Brill's tagger is that the rules are not hand-crafted, but are discovered utilizing the corpus given. A set of templates of rules is provided to the model using which it can find new features.

6.4.2   STOCHASTIC POS TAGGER

The term 'stochastic tagger' can allude to any number of various ways to deal with the issue of POS labeling. Any model which by one way or another consolidates recurrence or likelihood might be legitimately marked stochastic.

The least difficult stochastic taggers disambiguate words dependent on the likelihood that a word happens with a specific tag. At the end of the day, the tag experienced most every now and again in the training set with the word is the one allotted to an ambiguous occurrence of that word. The issue with this methodology is that while it might yield a valid tag for a given word, it can likewise yield prohibited successions of tags.

An option in contrast to the word recurrence approach is to figure the likelihood of a given arrangement of labels happening. This is in some cases alluded to as the n-gram approach, alluding to the way that the best tag for a given word is controlled by the likelihood that it happens with the n past labels. This methodology bodes well than the one characterized previously, in light of the fact that it considers the labels for individual words dependent on setting.

The following dimension of multifaceted nature that can be brought into a stochastic tagger consolidates the past two methodologies, utilizing both label grouping probabilities and word recurrence estimations. This is known as the HMM.

6.5   USE OF HIDDEN MARKOV MODEL (HMM) IN POS TAGGING

Before continuing with what is a HMM [8], let us first see what a Markov Model is. That will better help comprehend the significance of the term Hidden in HMMs.

6.5.1   AN INSIGHT INTO BASIC MARKOV MODEL

Markov concepts and hidden Markov can be quite some task to grasp considering the amount of mathematical support behind its intuition, to understand it and building a sound base may take quite some time and effort. Two machine learning enthusiasts, Sachin Malhotra (https://medium. com/@sachinmalhotra) and Divya Godayal (https://medium.com/@

divyagodayal), have beautifully described the basics of Markov model and importance of Hidden Markov in POS tagging in there blog posts (https://medium.freecodecamp.org/an-introduction-to-part-of-speech-tagging-and-the-hidden-markov-model-953d45338f24, https://medium.freecodecamp.org/a-deep-dive-into-part-of-speech-tagging-using-viterbi-algorithm-17c8de32e8bc). One can refer to have a sound idea on these topics yet not getting taken aback by the complexity of the same. The following discussion is inspired from their posts.

Let us assume there are just three types of climate conditions, in particular: Rainy, Sunny, and Cloudy. A little boy, Peter, wants to play outside. He loves sunny days because most of his friends are available on that day to play with him. Every day, his mom watches the climate in the first part of the day (that is the time at which Peter normally goes out to play). Once Peter wakes up in the morning, he comes up to her mom and enquires about how the weather would be like that day. Since she is a capable parent, she needs to answer that question as precisely as could be expected under the circumstances. In any case, the main thing she has is the past few days' data regarding the weather and based on that she needs to make her best guess about what it is going to be like that particular day.

Let's assume you have a succession of past few days' climate information (Figure 6.1).

Something like this: *Sunny, Rainy, Cloudy, Cloudy, Sunny, Sunny, Sunny, and Rainy.*

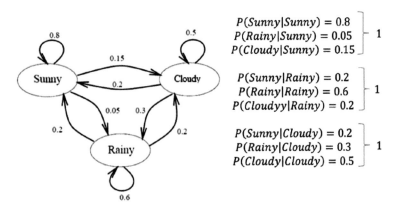

$P(Sunny|Sunny) = 0.8$
$P(Rainy|Sunny) = 0.05$ 1
$P(Cloudy|Sunny) = 0.15$

$P(Sunny|Rainy) = 0.2$
$P(Rainy|Rainy) = 0.6$ 1
$P(Cloudyy|Rainy) = 0.2$

$P(Sunny|Cloudy) = 0.2$
$P(Rainy|Cloudy) = 0.3$ 1
$P(Cloudy|Cloudy) = 0.5$

FIGURE 6.1   Climate state diagram. Reprinted with permission from Ref. 15.

In this way, the climate for some random day can be in any of the three states.

Suppose we choose to utilize a Markov Chain Model to take care of this issue. Presently utilizing the information that we have, we can build the accompanying state transition graph with the named probabilities as in Figure 6.1. So as to get the likelihood of the present climate given N past perceptions, we will utilize the Markovian Property.

$$P(q_1.q_n) = \prod^n_{i=1} P(q_i|q_{i-1}) \tag{1}$$

The Markov property proposes that the distribution for a random variable in the future depends entirely just on its appropriation in the present state, and none of the past states have any effect on the future states.

Let us look at an example just to perceive how the likelihood of the present state can be processed utilizing the formula above, considering the Markovian Property.

○ Exercise 1: Given that today is Sunny, what's the probability that tomorrow is Sunny and the next day Rainy?

$P(q_2, q_3|q_1) = P(q_2|q_1)P(q_3|q_1, q_2)$

$= P(q_2|q_1)\ P(q_3|q_2)$
$= P(Sunny|Sunny)\ P(Rainy|Sunny)$
$= (0.8)(0.05)$
$= 0.04$

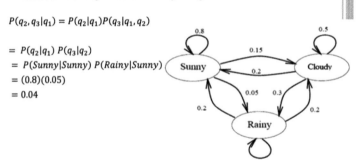

○ Exercise 2: Assume that yesterday's weather was Rainy, and today is Cloudy, what is the probability that tomorrow will be Sunny?

$P(q_3|q_1, q_2) = P(q_3|q_2)$

$= P(Sunny|Cloudy)$

$= 0.2$

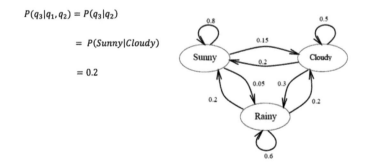

We can clearly see that as per the Markov property, the probability of tomorrow's weather being Sunny depends solely on today's weather and not on yesterday's.

Let us now proceed and see what are HMMs, and how it is different from a simple Markov Model.

6.5.2 HIDDEN MARKOV MODEL (HMM)

Let us consider you are the guardian of Peter. A standout amongst the most vital errands for you is to tuck Peter into bed and ensure he is sound sleeping. When you've tucked him in, you need to ensure he's sleeping in reality and not up to some naughtiness. You can't, nonetheless, go into the room to ensure that, as that would doubtlessly wake Peter up. So all you need to choose are the sounds that may originate from the room. Either the room is *quiet* or there is *noise* originating from the room. These are your states.

Peter's mom has given you the below state chart having a few states, observations, and probabilities (Figure 6.2).

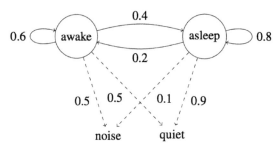

FIGURE 6.2   State chart of the sleeping baby problem. Reprinted with permission from Ref. 15.

Note that there is no immediate connection between sound from the room and Peter being sleeping.

There are two sorts of probabilities that we can see from the statechart.

1. One is the *Emission probabilities*, which speak of the probabilities of mentioning certain objective facts given a specific state. For instance, we have P(noise | awake) = 0.5.
2. The other being *Transition probabilities*, which speak of the likelihood of progressing to another state given a specific state. For instance, we have P(asleep | wakeful) = 0.4.

The Markov property, as would be appropriate to the model we have considered here, would be that the likelihood of Peter being in a state depends *only* on the past state.

But there is an unmistakable blemish in the Markov property. On the off chance that Peter has been wakeful for 60 minutes, at that point, the likelihood of him dozing off is higher than it has been conscious for only 5 minutes. Along these lines, history matters. Hence, the Markov state machine-based model isn't totally right. The Markov property, albeit wrong, makes this issue entirely tractable.

We, for the most part, watch longer stretches of the youngster being alert and being sleeping. In the event that Peter is wakeful now, the likelihood of him remaining alert is higher than of him resting. Consequently, 0.6 and 0.4 in the above diagram (Tables 6.1 and 6.2).

P(awake | awake) = 0.6 and P(asleep | awake) = 0.4

TABLE 6.1   The Transition Probabilities Matrix

	Awake	Asleep
Awake	0.6	0.4
Asleep	0.2	0.8

TABLE 6.2   The Emission Probabilities Matrix

	Noise	Quiet
Awake	0.5	0.5
Asleep	0.1	0.9

Before really attempting to take care of the current issue utilizing HMMs, let's understand how we can relate this situation to the task of Part of Speech Tagging. We realize that to show any issue utilizing a HMM we need a lot of perceptions and a lot of conceivable states. The states in a HMM are covered up.

In the parts of speech labeling problem, the *observations* are simply the *words* in the given grouping. The *POS labels* for the words resemble the *'states'* which are hidden.

The transition probabilities would be to some degree like P(VP | NP) that is, what is the likelihood of the present word having a tag of Verb Phrase given that the past tag was a Noun Phrase. Emission probabilities would be P(john | NP) or P(will | VP) that is, what is the likelihood that the word is, state, John given that the tag is a Noun Phrase.

6.5.3   APPLICATION OF VITERBI ALGORITHM IN POS TAGGING

Our problem in hand can be formulated in as follows:

Given the state graph and a succession of N perceptions after some time, we have to tell the condition of the child at the present point in time. Numerically, we have N perceptions over occasion's t0, t1, t2. tN. We need to see whether Peter would be alert or snoozing, or rather which state is increasingly plausible at time tN+1.

Before proceeding onward to the Viterbi Algorithm, let's understand through a point by point clarification of how the POS labeling issue can be addressed utilizing HMMs [9].

6.5.3.1   GENERATIVE MODELS AND THE NOISY CHANNEL MODEL

A great deal of issues in natural language processing (NLP) is tackled utilizing a supervised learning approach.

Supervised tasks in machine learning are characterized as follows. We take into consideration training data $(x_{(1)}, y_{(1)}).(x_{(m)}, y_{(m)})$, where every instance comprises of an information $x_{(i)}$ matched with a name $y_{(i)}$. We use X to allude to the arrangement of input data, and Y to allude to the possible labels. Our assignment is to train a function f: $X \rightarrow Y$ that maps any input x to a corresponding label f(x).

In labeling issues, each $x_{(i)}$ would be an arrangement of words $X_1 X_2 X_3....X_{n(i)}$, and each $y_{(i)}$ would be a succession of labels $Y_1 Y_2 Y_3 ... Y_{n(i)}$ (we use n(i) to allude to the length of the i'th training data). X would allude to the arrangement of all successions $x_1. x_n$, and Y would be the arrangement of all label successions $y_1. y_n$. Our goal would be to learn a function f: $X \rightarrow Y$ that would maps sentences to label (tag) sequences.

The idea of conditional probability may come handy to estimate this problem here. p(y | x) which is the likelihood of the yield y given an information x. The parameters of the model would be assessed utilizing the training examples. Now, when we give and unknown input x, we would expect to get:

$$f(x) = argmax(p(y \mid x)) \; \forall y \in Y \qquad (2)$$

So, given the training data, the above is a conditional model to solve this generic problem. Another methodology that is popularly used in ML and NLP is to utilize a generative model. Unlike in conditional model where we evaluate directly the conditional probability p(y|x), in generative models, we rather show the joint likelihood p(x, y) over all the (x, y) sets.

We can additionally break down the joint likelihood into more straight forward values utilizing Bayes' rule:

$$p(x,y) = p(y)p(x|y) \qquad (3)$$

where; p(y) signifies the prior probability of any input with the label y and p(x | y) is the conditional probability of input x given the label y.

We can utilize this decomposition and the Bayes principle to decide the conditional probability.

$$p(y|x) = p(y)p(x|y)/p(x) \qquad (4)$$

Keep in mind that, we needed to evaluate the function:

$$f(x) = argmax(p(y|x)) \; \forall y \in Y \qquad (5)$$
$$f(x) = argmax(p(y) * p(x | y)) \qquad (6)$$

The reason we avoided the denominator here is on the grounds that the likelihood p(x) stays constant with the changing output labels. Thus, from a computational point of view, it is treated as normalization constant and is usually disregarded.

Models that breaks a joint likelihood into terms p(y) and p(x|y) are popularly known as *noisy-channel models*. Generally, when we come across a test instance x, we assume it was produced by the following two steps.

1. A label y has been picked with likelihood p(y)
2. The model x has been produced from the distribution p(x|y). You can think in this way that the model p(x|y) accepts label y as input and made to deliver x as its output.

6.5.3.2 GENERATIVE PARTS OF SPEECH TAGGING MODEL

Let's consider a vocabulary of words V and a limited sequence of labels/tags K. Set S will be the set of all sequence, tags pairs $<x_1, x_2, x_3. x_n, y_1, y_2, y_3., y_n>$ where n > 0 $\forall x \in V$ and $\forall y \in K$.

A generative tagging model is then the one where for any $<x_1.x_n, y_1.y_n> \in S$,

$$p(x_1.x_n, y_1.y_n) > = 0 \tag{7}$$

$$\sum_{<x1.xn,y1.yn> \in S} p(x_1.x_n, y_1.y_n) = 1 \tag{8}$$

Given a generative tagging model, the function that we talked about earlier from input to output becomes:

$$f(x_1.x_n) = \arg\max_{y1.yn} p(x_1.x_n, y_1.y_n) \tag{9}$$

In this way for some random sequence of words as input, the output is the maximum likelihood tag sequence from the model. Having characterized the generative model, we have to make sense of three distinct things:

1. How precisely do we characterize the generative model likelihood $p(<x_1, x_2, x_3. x_n, y_1, y_2, y_3., y_n>)$;
2. How would we gauge the parameters of the model; and
3. How would we effectively compute.

$$f(x_1.x_n) = \arg\max_{y1.yn} p(x_1.x_n, y_1.y_n) \tag{10}$$

Now, based on our problem in hand of POS tagging and the one where Peter is involved, let's try to address the above three questions.

6.5.3.3   DESCRIBING THE GENERATIVE MODEL

Let us first try to survey the probability $p(x_1. x_n, y_1.y_n)$ using the HMM. We can have any N-gram HMM which considers events in the past window of size N. The following segment relates with a Trigram HMM.

6.5.3.3.1   Trigram HMM

A trigram HMM [10] can be characterized utilizing:

- A limited number of states.
- An observation sequence.
- $q(s|u,v)$: Transition probability is characterized as the likelihood of a state "s" seeming to occur in the wake of watching "u" and "v" in the observation sequence.
- $e(x|s)$: Emission probability is characterized as the likelihood of mentioning an objective fact 'x' given that the state was 's.'

At that point, the generative model likelihood would be evaluated as:

$$p(x_1.x_n,y_1.y_{n+1}) = \prod^{n+1}_{i=1} q(y_i|y_{i-2}, y_{i-1})\prod^n_{i=1}e(x_i|y_i) \qquad (11)$$

With respect to the 'Sleeping-baby problem' that we are thinking about, we will have just two conceivable states: that the infant is either awake or he is sleeping. The overseer can mention just two observations after some time. Either there is some sound of commotion coming from the room or the room is completely calm. The below representation shows a possible depiction of the sequence of observations and states (Figure 6.3).

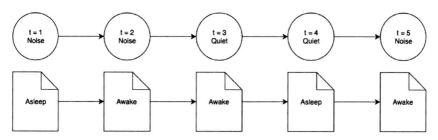

FIGURE 6.3 Observations and states over an interval of time for the sleeping-baby problem. Reprinted with permission from Ref. 16.

In the POS tagging problem, real-tag values allotted to words will be referred to as the states and observations will be the words. In a HMM, the states remain hidden always, the states being the only visible entities to us. On line with this, we can represent the states and observations for the POS tagging problem as follows (Figure 6.4).

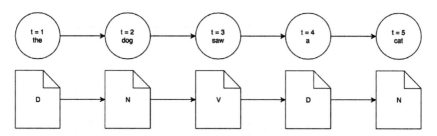

FIGURE 6.4 Observations and states over an interval of time for the POS labeling problem. Reprinted with permission from Ref. 16.

6.5.3.3.2   *Estimating the Model's Parameters*

Assuming to have access to some training data, containing set of examples, each representing a sequence of observations and each observation is associated with a state. Now, the question is how to estimate the parameters of the model based on the above-given data. Next, we estimate the model's parameters by reading various counts off of the training corpus we have and then computing maximum likelihood estimates. The transition probability and emission probability for a Trigram HMM is calculated as follows:

$$q(s|u,v) = c(u,v,s)/c(u,v)$$
$$e(x|s) = c(s\text{->}x)/c(s)$$

We already know that the first term represents transition probability and the second term represents the emission probability. Let us look at what the four different counts mean in the terms above.

c(u, v, s) represents the trigram count of states u, v, and s. Meaning it represents the number of times the three states u, v, and s occurred together in that order in the training corpus.

c(u, v) following along similar lines as that of the trigram count, this is the bigram count of states u and v given the training corpus.

c(s → x) is the number of times in the training set that the state s and observation x are paired with each other. And finally,

c(s) is the prior probability of an observation being labeled as the state s.

Let us look at a sample training set for the toy problem first and see the calculations for transition and emission probabilities using the same.

The BLUE markings represent the transition probability, and RED is for emission probability calculations.

Note that since the example problem only has two distinct states and two distinct observations, and given that the training set is very small, the calculations shown below for the example problem are using a bigram HMM instead of a trigram HMM.

Peter's mother was maintaining a record of observations and states. And thus she even provided you with a training corpus to help you get the transition and emission probabilities.

- **Transition Probability Example:**
  The Training Corpus for transition probability:

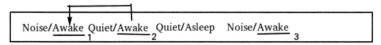

Noise/Awake Quiet/Awake Quiet/Asleep Noise/Awake
      1   2   3

The calculations for awake appearing after Awake will be:

$$q(Awake|Awake) = \frac{c(Awake|Awake) \longleftarrow \text{Count of Bigram}}{c(Awake) \longleftarrow \text{Count of tag}}$$

$$= 1/3$$

- **Emission Probability Example:**
  The Training corpus for emission probability be like:

    Noise/Awake Quiet/Awake Quiet/Asleep Noise/Awake

The calculations for observing 'Quiet' when the state is 'Awake' will be:

$$c(Quiet|Awake) = \frac{c(Quiet \rightarrow Awake) \longleftarrow \text{Observation}}{c(Awake) \longleftarrow \text{State}}$$

$$= 1/3$$

That was quite simple, since the training set was very small. Let us look at a sample training set for our actual problem of part of speech tagging. Here we can consider a trigram HMM, and we will show the calculations accordingly.

We will use the following sentences as a corpus of training data (the notation word/TAG means word tagged with a specific part-of-speech tag).

> eat/VB breakfast/NN at/IN morning/NN time/NN
> take/VB time/NN with/IN arrow/NN projects/NN
> horse/NN riders/NN like/VB the/DT airport/NN
> paper/NN flies/VB on/IN hydrogen/NN gas/NN
> bees/NN sting/VB like/IN some/DT flies/NN
> beans/NN soil/VB an/Dt iron/NN grill/NN
> flies/NN smell/VB and/DT arrow/NN drink/NN
> people/NN like/VB an/DT army/NN arrow/NN
> dinner/NN time/NN flies/VB all/DT day/NN
> horse/NN flies/NN time/VB morning/NN rays/NN

The training set that we have is a tagged corpus of sentences. Every sentence consists of words tagged with their corresponding part of speech tags. E.g.: eat/VB means that the word is "eat" and the part of speech tag in this sentence in this very context is "VB" i.e., Verb Phrase. Let us look at a sample calculation for transition probability and emission probability just like we saw for the baby sleeping problem.

1. **Transition Probability:** Let's say we want to calculate the transition probability q(IN | VB, NN). For this, we see how many times we see a trigram (VB, NN, IN) in the training corpus in that specific order. We then divide it by the total number of times we see the bigram (VB, NN) in the corpus.

2. **Emission Probability:** Let's say we want to find out the emission probability e(an | DT). For this, we see how many times the word "an" is tagged as "DT" in the corpus and divide it by the total number of times we see the tag "DT" in the corpus.

$$q(IN \mid VB,NN) = \frac{c(VB,NN,IN)}{c(VB,NN)} \qquad \begin{array}{l} \text{count of trigram} \\ \text{count of bigram} \end{array}$$

$$= 2/3$$

$$e(an \mid DT) = \frac{c(an \rightarrow DT)}{c(DT)} \qquad \begin{array}{l} \text{state /tag} \\ \text{Observation Word} \end{array}$$

$$= 3/6$$

So if you look at these calculations, it shows that calculating the model's parameters is not computationally expensive. That is, we don't have to do multiple passes over the training data to calculate these parameters. All we need are a bunch of different counts, and a single pass over the training corpus should provide us with that.

Moving on, let's look at the final step that we need to look at given a generative model. That step is efficiently calculating:

$$f(x_1.x_n) = \arg\max_{y1.yn} p(x_1.x_n, y_1.y_n) \qquad (12)$$

We will be looking at the famous *Viterbi Algorithm* for this calculation.

Objective of Viterbi algorithm: To find the most likely hidden state sequence, given a sequence of observations.

Finally, we are going to solve the problem of finding the most likely sequence of labels given a set of observations x1 ... xn. That is, we are to find out

$$\arg\max{}_{y1.yn+1} p(x_1.x_n, y_1.y_{n+1}) \tag{13}$$

As discussed earlier, we define the probability of a sequence of labels, given a sequence of observations over "n" time steps, in terms of transition and emission probabilities.

$$p(x_1.x_n, y_1.y_{n+1}) = \prod{}^{n+1}_{i=1} q(y_i|y_{i-2}, y_{i-1}) \prod{}^{n}_{i=1} e(x_i|y_i) \tag{14}$$

Before looking at an optimized algorithm to solve this problem, let us first look at a simple brute force approach to this problem. Basically, we need to find out the most probable label sequence given a set of observations out of a finite set of possible sequences of labels. Let's look at the total possible number of sequences for a small example for our example problem and also for a part of speech tagging problem.

Say we have the following set of observations for the example problem.

Noise Quiet Noise

We have two possible labels *{Asleep and Awake}*. Some of the possible sequence of labels for the observations above are:

Awake Awake Awake
Awake Awake Asleep
Awake Asleep Awake
Awake Asleep Asleep

In all, we can have $2^3 = 8$ possible sequences. This might not seem like very many, but if we increase the number of observations over time, the number of sequences would increase exponentially. This is the case when we only had two possible labels. What if we have more? As is the case with part of speech tagging.

So the exponential growth in the number of sequences implies that for any sentence of average length, the brute force approach would not work out as it would take too much time to execute.

An efficient alternative to this brute force approach is Viterbi Algorithm [11], a dynamic programming algorithm that can find the highest probable tag sequence.

Let us first define some terms that would be useful in defining the algorithm itself. We already know that the probability of a label sequence given a set of observations can be defined in terms of the transition probability and the emission probability. Mathematically, it is

$$p(x_1.x_n,y_1.y_{n+1}) = \prod_{i=1}^{n+1} q(y_i|y_{i-2}, y_{i-1})\prod_{i=1}^{n} e(x_i|y_i) \tag{15}$$

Let us look at a truncated version of this which is

$$r(y_1.y_k) = \prod_{i=1}^{k} q(y_i|y_{i-2}, y_{i-1})\prod_{i=1}^{k} e(x_i|y_i) \tag{16}$$

and let us call this the cost of a sequence of length k.

So the definition of "r" is simply considering the first k terms off of the definition of probability where $k \in \{1.n\}$ and for any label sequence $y_1...y_k$.

Next, we have the set $S(k, u, v)$ which is basically the set of all label sequences of length k that end with the bigram (u, v) i.e., set of sequences $y_1. y_k$ such that $y_{k-1} = u$, $y_k = v$.

Finally, we define the term $\pi(k, u, v)$ which is basically the sequence with the maximum cost.

$$\prod (k,u,v) = \max_{<y1.yk>\in S(k,u,v)} r(y1.yk) \tag{17}$$

The main idea behind the Viterbi Algorithm is that we can calculate the values of the term $\pi(k, u, v)$ efficiently in a recursive, memoized fashion. In order to define the algorithm recursively, let us look at the base cases for the recursion.

$$\pi (0, *, *) = 1$$
$$\pi (0, u, v) = 0$$

Since we are considering a trigram HMM, we would be considering all of the trigrams as a part of the execution of the Viterbi Algorithm.

Now, we can start the first trigram window from the first three words of the sentence but then the model would miss out on those trigrams where the first word or the first two words occurred independently. For that reason, we consider two special start symbols as * and so our sentence becomes:

$$* * x_1 x_2 x_3. xn$$

And the first trigram we consider then would be $(*, *, x_1)$ and the second one would be $(*, x_1, x_2)$.

Now that we have all our terms in place, we can finally look at the recursive definition of the algorithm which is basically the heart of the algorithm.

This definition is clearly recursive, because we are trying to calculate one π term and we are using another one with a lower value of k in the recurrence relation for it.

$$\max\nolimits_{y1.yn+1} p(x_1.x_n,y_1.y_{n+1}) = \max\nolimits_{u\in K,\, v\in K}(\textstyle\prod(n, u, v) \times q(STOP|u,v)) \quad (18)$$

Every sequence would end with a special STOP symbol. For the trigram model, we would also have two special start symbols "*" in the beginning.

Have a look at the pseudo-code for the entire algorithm.

Input: a sentence $x_1 \ldots x_n$, parameters $q(s|u,v)$ and $e(x|s)$.
Initialization: Set $\pi(0, *, *) = 1$, and $\pi(0, u, v) = 0$ for all (u, v) such that $u \neq *$ or $v \neq *$.
Algorithm:

- For $k = 1 \ldots n$,

  - For $u \in \mathcal{K}, v \in \mathcal{K}$,

  $$\pi(k, u, v) = \max_{w\in\mathcal{K}} (\pi(k - 1, w, u) \times q(v|w, u) \times e(x_k|v))$$

- **Return** $\max_{u\in\mathcal{K},v\in\mathcal{K}} (\pi(n, u, v) \times q(STOP|u, v))$

At first, the algorithm recursively fills in the π(k, u, v) values. It then uses the identity described before to calculate the highest probability for any sequence. $O(n|K|^3)$ is the running time for the algorithm, from which it can be inferred that it is linear in the length of the sequence, and cubic with respect to the tags frequency (tags count).

Note: We would be showing calculations for the baby sleeping problem and the part of speech tagging problem based off a bigram HMM only [12]. The calculations for the trigram are left to the reader to do themselves. But the code that is attached at the end of this article is based on a trigram HMM. It's just that the calculations are easier to explain and portray for the Viterbi algorithm when considering a bigram HMM instead of a trigram HMM.

The recursive formula based on a bigram HMM is extremely similar to the one we saw before for the trigram model, except that now we are only concerning ourselves with the current label and the one before, instead of two before. The complexity of the algorithm now becomes $O(n|K|^2)$.

6.5.3.3.3   Calculations for the Baby Sleeping Problem

Note that when we are at this step, that is, the calculations for the Viterbi Algorithm to find the most probable tag sequence given a set of observations over a series of time steps, we assume that transition and emission probabilities have already been calculated from the given corpus. Let's have a look at a sample of transition and emission probabilities for the baby-sleeping problem that we would use for our calculations of the algorithm.

The baby starts by being awake, and remains in the room for three-time points, t_1, ..., t_3 (three iterations of the Markov chain). The observations are: quiet, quiet, and noise. Have a look at the following diagram that shows the calculations for up to two time-steps. The complete diagram with all the final set of values will be shown afterwards (Figure 6.5).

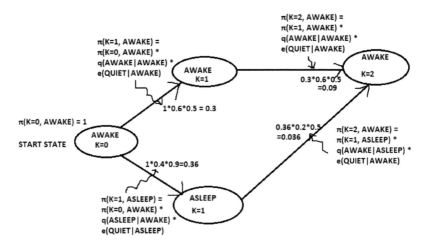

FIGURE 6.5 Transition and emission probabilities calculations. Reprinted with permission from Ref. 16.

We have not shown the calculations for the state of "asleep" at k = 2 and the calculations for k = 3 in the above diagram to keep things simple.

Now that we have all these calculations in place, we want to calculate the most likely sequence of states that the baby can be in over the different given time steps. So, for k = 2 and the state of Awake, we want to know the most likely state at k = 1 that transitioned to Awake at k = 2 (k = 2 represents a sequence of states of length 3 starting off from 0 and t = 2 would mean the state at time-step 2. We are given the state at t = 0, i.e., Awake) (Figure 6.6).

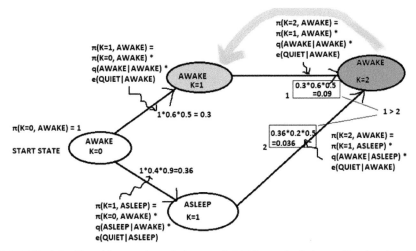

FIGURE 6.6 Transition and emission probabilities calculations. Reprinted with permission from Ref. 16.

Clearly, if the state at time-step 2 was AWAKE, then the state at time-step 1 would have been AWAKE as well, as the calculations point out. So, the Viterbi Algorithm not only helps us find the π(k) values, that is the cost values for all the sequences using the concept of dynamic programming, but it also helps us to find the most likely tag sequence given a start state and a sequence of observations.

6.5.3.3.4   Calculations for the Parts of Speech Tagging Problem

Let us look at a slightly bigger corpus for the POS tagging and the corresponding Viterbi graph showing the calculations and back-pointers for the Viterbi Algorithm.

Here is the corpus that we will consider:

eat/VB breakfast/NN at/IN morning/NN time/NN
take/VB time/NN with/IN arrow/NN projects/NN
horse/NN riders/NN like/VB the/DT airport/NN
paper/NN flies/VB on/IN hydrogen/NN gas/NN
bees/NN sting/VB like/IN some/DT flies/NN
beans/NN soil/VB an/DT iron/NN grill/NN
flies/NN smell/VB an/DT arrow/NN drink/NN
people/NN like/VB an/DT army/NN arrow/NN
dinner/NN time/NN flies/VB all/DT day/NN
horse/NN flies/NN time/VB morning/NN rays/NN

Now take a look at the transition probabilities calculated from this corpus (Figure 6.7).

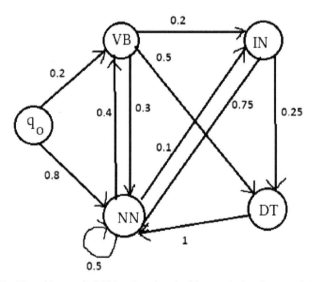

FIGURE 6.7   Transition probabilities. Reprinted with permission from Ref. 16.

Here, $q_0 \rightarrow$ VB represents the probability of a sentence starting off with the tag VB that is the first word of a sentence being tagged as VB. Similarly, $q_0 \rightarrow$ NN represents the probability of a sentence starting with the tag NN. Notice that out of 10 sentences in the corpus, 8 starts with NN and 2 with VB and hence the corresponding transition probabilities.

As for the emission probabilities, ideally, we should be looking at all the combinations of tags and words in the corpus. Since that would be too much, we will only consider emission probabilities for the sentence that would be used in the calculations for the Viterbi Algorithm.

Time flies like an arrow

The emission probabilities for the sentence above are given in Table 6.3.

TABLE 6.3 Emission Probabilities

	time	flies	like	an	arrow
VB	1/10 = 0.1	2/10 = 0.2	2/10 = 0.2	0	0
NN	3/30 = 0.1	3/30 = 0.1	0	0	3/30 = 0.1
IN	0	0	¼ = 0.25	0	0
DT	0	0	0	3/6 = 0.5	0

Finally, we are ready to see the calculations for the given sentence, transition probabilities, emission probabilities, and the given corpus.

Most likely sequence: *NN, NN, VB, DT, NN.*

Dashed lines denote zero probability transitions (Figure 6.8).

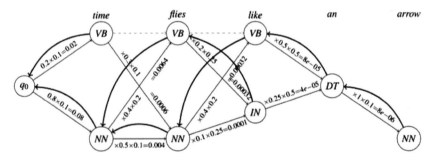

FIGURE 6.8 Transmission and emission probabilities. Reprinted with permission from Ref. 16.

The bucket below each word is filled with the possible tags seen next to the word in the training corpus. The given sentence can have the combinations of tags depending on which path we take which is all combinations of sequence paths (Figure 6.9).

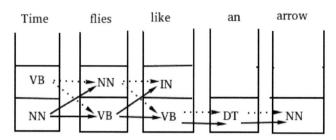

FIGURE 6.9   All combinations of sequence paths. Reprinted with permission from Ref. 16.

There might be some path in the computation graph for which we do not have a transition probability. So our algorithm can just discard that path and take the other path (Figure 6.10).

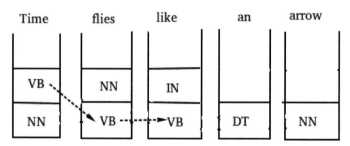

FIGURE 6.10   Path with no transition probability. Reprinted with permission from Ref. 16.

In the above diagram, we discard the path marked by a dotted line since we do not have q(VB|VB). The training corpus never has a VB followed by VB. So in the Viterbi calculations, we end up taking q(VB|VB) = 0 and a single 0 in the calculations would make the entire probability or the maximum cost for a sequence of tags/labels to be 0. This, however, means that we are ignoring the combinations which are not seen in the training corpus. The corpus that we considered here was very small. Consider any reasonably sized corpus with a lot of words and we have a major problem of the sparsity of data. To solve this problem of data sparsity, we resort to a solution called Smoothing.

6.6   POPULAR POS TAGGERS

Among all the tools available for parts-of-speech tagging, below we mention some frequently used POS taggers.

6.6.1 NLTK

The natural language toolkit (NLTK) is heavily used for building programs for text analysis. NLTK's POS tagger (nltk.pos_tag()) is one of the most popular POS tagger that is being used extensively in varied applications in NLP.

NLTK's POS tagger tags each token or word in any of the below mentioned POS tags (Table 6.4):

TABLE 6.4   List of POS Tags in NLTK POS Tagset

CC: coordinating conjunction	PRP$: possessive pronoun
CD: cardinal digit	RB: adverb
DT: determiner	RBR: adverb, comparative
EX: existential there	RBS: adverb, superlative
FW: foreign word	RP: particle
IN: preposition/subordinating conjunction	TO: to
JJ: adjective	UH: interjection
JJR: adjective	VB: verb, base form
JJS: adjective	VBD: verb, past tense
LS: list marker	VBG: verb, gerund/present participle
MD: modal	VBN: verb, past participle
NN: noun, singular	VBP: verb, sing. present, non-3d
NNS: noun plural	VBZ: verb, 3rd person sing. present
NNP: proper noun	WDT: wh-determiner
NNPS: proper noun	WP: wh-pronoun
PDT: predeterminer	WP$: possessive wh-pronoun
POS: possessive ending	WRB: wh-adverb
PRP: personal pronoun	

For instance, consider the following sentence:
Automatic Parts of Speech tagging is so cool!

- **Python Code Snippet:**
  ```
  >>> import nltk
  >>> tokens = nltk.word_tokenize('Automatic Parts of Speech tagging is so cool!')
  >>>nltk.pos_tag(tokens)
  ```

Output:
[('Automatic,' 'JJ'), ('Parts,' 'NNS'), ('of,' 'IN'), ('Speech,' 'NNP'), ('tagging,' 'NN'), ('is,' 'VBZ'), ('so,' 'RB'), ('cool,' 'JJ'), ('!,'.')]

6.6.2 PYSTATPARSER

pyStatParser is a simple python statistical parser that returns NLTK parse Trees. It comes with public treebanks (QuestionBank and Penn Tree Bank) and it generates the grammar model only the first time you instantiate a Parser object. It uses a CKY algorithm and it parses average length sentences in under a second.

For instance, consider the following sentence:

How can the net amount of entropy of the universe be massively decreased?

- **Python Code Snippet:**
  >>> from stat_parser import Parser
  >>> parser = Parser()
  >>>print(parser.parse("How can the net amount of entropy of the universe be massively decreased?"))

Output:
(SBARQ
(WHADVP (WRB how))
(SQ
(MD can)
(NP
(NP (DT the) (JJ net) (NN amount))
(PP
(IN of)
(NP
(NP (NNS entropy))
(PP (IN of) (NP (DT the) (NN universe))))))))
(VP (VB be) (ADJP (RB massively) (VBN decreased))))
(. ?))

6.6.3 STANFORD POS

Stanford is probably considered the most widely used POS tagger.
For instance, consider the following sentence:

What in the earth are parts of speech tags?

- **Python Code Snippet:**
  from nltk.tag.stanford import StanfordTagger

sample = 'What in the earth are parts of speech tags?';
tok = sample.split()
model_file = '/nlp/stanford-postagger/models/bidirectional-distsim-wsj-0-18.tagger'
model_jar = '/nlp/stanford-postagger/stanford-postagger.jar'
model_tagger = StanfordTagger(model_file, path_to_jar = model_jar, encoding = 'UTF-8')
print(model_tagger.tag(tok))

Output:

[(u'What,'u'WP'), (u'in,'u'IN'), (u'the,'u'DT'), (u'earth,'u'NN'),
(u'are,' u'VBP'), (u'parts,' u'NNS'), (u'of,' u'IN'), (u'speech,'
u'NN'), (u'tags,'u'NNS'), (u'?,'u.")]

6.6.4 MALTPARSER

MaltParser is a framework for information-driven dependency parsing, which can be utilized to incite a parsing model from treebank information and to parse new information utilizing an actuated model. MaltParser can be portrayed as an information-driven parser-generator. While a conventional parser-generator builds a parser given a language, an information-driven parser-generator develops a parser given a treebank. MaltParser is an execution of inductive dependency parsing, where the syntactic examination of a sentence adds up to the inference of a dependency structure, and where inductive machine learning is utilized to manage the parser at non-deterministic choice points [14].

- **Python Code Snippet:**
  You need to download and extract the Malt Parser and pre-trained model (*engmalt.poly-1.7.mco*) before using the parser.
  >>> import os
  >>>mp = nltk.parse.malt.MaltParser(os.getcwd(), model_filename = "engmalt.poly-1.7.mco")
  >>> sentence = 'Yet another Parts of Speech tagger. I am loving it.'
  >>> print(mp.parse_one(sentence.split()).tree())

Output:

(it. (tagger. Yet another Parts of Speech) I am loving)

6.6.5   PATTERN

The 'pattern.en' has a tag method that can be used for POS Tagging for English content, the POS tags follows Penn Treebank II tag set. pattern. en is English-specific NLTK toolkit. Because language is ambiguous (e.g., I can ↔ a can) it uses statistical approaches + regular expressions. This means that it is fast, quite accurate and occasionally incorrect. It has a part-of-speech tagger that identifies word types (e.g., noun, verb, adjective), word inflection (conjugation, singularization) and a WordNet API.

- **Python Code Snippet:**

Tagging

```
>>> from pattern.en import tag
>>> sentence = 'Indeed parts of speech tagging is quite important
in solving various natural language processing tasks.'
>>> tag(sentence)
```

Output:

[('Indeed,' 'RB'), ('parts,' 'NNS'), ('of,' 'IN'), ('speech,' 'NN'), ('tagging,' 'VBG'), ('is,' 'VBZ'), ('quite,' 'RB'), ('important,' 'JJ'), ('in,' 'IN'), ('solving,' 'VBG'), ('various,' 'JJ'), ('natural,' 'JJ'), ('language,' 'NN'), ('processing,' 'VBG'), ('tasks,' 'NNS'), (.,'.')]

Parsing

```
>>> from pattern.en import parse
>>> sentence = 'Indeed parts of speech tagging is quite important
in solving various natural language processing tasks.'
>>>parse(sentence, relations = True, lemmata = True)
```

Output:

'Indeed/RB/B-ADVP/O/O/indeed parts/NNS/B-NP/O/O/part of/IN/ B-PP/B-PNP/O/of speech/NN/B-NP/I-PNP/NP-SBJ-1/speech tagging/VBG/B-VP/I-PNP/VP-1/tag is/VBZ/I-VP/O/VP-1/be quite/RB/B- ADJP/O/O/quite important/JJ/I-ADJP/O/O/important in/IN/B-PP/B- PNP/O/in solving/VBG/B-VP/I-PNP/VP-2/solve various/JJ/B-NP/I- PNP/NP-OBJ-2*NP-SBJ-3/various natural/JJ/I-NP/I-PNP/NP-OBJ- 2*NP-SBJ-3/natural language/NN/I-NP/I-PNP/NP- OBJ-2*NP-SBJ-3/ languageprocessing/VBG/B-VP/I-PNP/VP-3/process tasks/NNS/B-NP/ I-PNP/NP-OBJ-3/task.//O/O/O/.'

The pattern has some very cool features beyond just POS. For instance, it comes with a search() method where you can find POS matching a rule in a parse tree. For example search ('VB*,' tree) matches even wildcard VB...verbs. Very useful for feature engineering tasks.

6.6.6   SPACY

Spacy is generally perceived as one of the amazing and propelled library used to actualize NLP tasks. spaCy is a free, open-source library for cutting edge NLP in Python. In case you're working with a great deal of data (text), you will, in the long run, need to find out about it. For instance, what's it about? What do the words mean in the setting? Who is doing what to whom? What organizations and items are referenced? Which writings are like one another? After tokenization, spaCy can parse and label a given Doc. This is the place the measurable model comes in, which empowers spaCy to make a forecast of which tag or mark undoubtedly applies in this unique circumstance. In the same way as other NLP libraries, spaCy encode all strings to hash values to diminish memory use and improve productivity. So to get the meaningful string portrayal of an attribute, we have to include an underscore_to its name.

- **Python Code Snippet:**

  import spacy
  test = spacy.load('en_core_web_sm')
  res = nlp(u 'Hi, I am Spacy. I can perform POS tagging, Tokenization, Lemmatizing, Dependency Parsing, Named Entity Representation, etc.')
  for token in res:
  print(token.text, token.lemma_, token.pos_, token.tag_, token.is_stop)

Output:

Hi hi INTJ UH False
, PUNCT, False
I-PRON-PRON PRP False
am be VERB VBP True
Spacy spacy PROPN NNP False
. PUNCT. False
I-PRON-PRON PRP False

can can VERB MD True
perform perform VERB VB False
POS pos PROPN NNP False
tagging tag VERB VBG False
, PUNCT, False
Tokenization tokenization PROPN NNP False
, PUNCT, False
Lemmatizing lemmatizing PROPN NNP False
, PUNCT, False
Dependency dependency PROPN NNP False
Parsing parsing PROPN NNP False
, PUNCT, False
Named named PROPN NNP False
Entity entity PROPN NNP False
Representation representation PROPN NNP False
, PUNCT, False
etc etc X FW False
. PUNCT. False

The below implementation uses spaCy's built-in displaCy visualizer to display how the sentence and its dependencies look like:

- **Python Code Snippet:**

```
import spacy
from spacy import displacy
test = spacy.load('en')
res = test(u'They ate the pizza with anchovies')
displacy.serve(res, style = 'dep')
```

Output:

A correct parse links "with" to "pizza", while an incorrect parse links "with" to "eat"

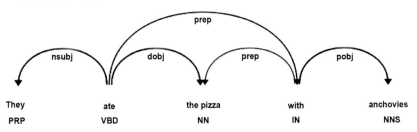

spaCy can be faster than the NLTK wrapped models or Stanford. It has a lot or richness to its functionality. Not always as pythonic but certainly the route to look for a large scale project where you know already what you want to do.

6.6.7   TEXTBLOB

TextBlob is a python library and offers a simple API to access its methods and perform basic NLP tasks. A good thing about TextBlob is that they are just like python strings. So, one can transform and play with it same like we did in python. TextBlob stands on the giant shoulders of NLTK and pattern, and plays nicely with both. If we compare execution speed among NLTK, spacy, and Textblobit cane arranged like this: Spacy >TextBlob> NLTK. Textblob is being used extensively in various NLP tasks such as classification, POS tagging, translation, etc.

For instance, consider the following sentence:

TextBlob is also a good parts of speech tagger.

- **Python Code Snippet:**

    >>> from textblob import TextBlob
    >>>test = TextBlob('TextBlob is also a good parts of speech tagger.')
    >>>test.parse()

Output:

'TextBlob/NN/B-NP/O is/VBZ/B-VP/O also/RB/B-ADVP/O a/DT/ B-NP/O good/JJ/I-NP/O parts/NNS/I-NP/O of/IN/B-PP/B-PNP speech/NN/B-NP/I-PNP tagger/NN/I-NP/I-PNP./.O/O'
    >>>blob.tags

Output:

[('TextBlob,' 'NNP'), ('is,' 'VBZ'), ('also,' 'RB'), ('a,' 'DT'), ('good,' 'JJ'), ('parts,' 'NNS'), ('of,' 'IN'), ('speech,' 'NN'), ('tagger,' 'NN')]

6.7   CONCLUSION

Identifying part of speech tags is much more complicated than simply mapping words to their part of speech tags. This is because POS tagging is not something that is generic. It is quite possible for a single word to have a different part of speech tag in different sentences based on different contexts. That is why it is impossible to have a generic mapping for POS tags. As you can see, it is not possible to manually find out different part-of-speech tags for a given corpus. New types of contexts and new words keep coming up in dictionaries in various languages, and manual POS tagging is not scalable in itself. That is why we rely on machine-based POS tagging. Parts of speech tagging can be important for syntactic and semantic analysis. The process of labeling each word with some parts-of-speech tag is known as POS tagging. POS tagging can be performed using both supervised as well as unsupervised approaches. This chapter focuses on supervised approaches (Rule-based and Stochastic-based). Under the Stochastic approach, an elaborate discussion on POS labeling using the Viterbi algorithm has been done here. We have also discussed various popularly used POS taggers along with their implementation with proper examples to give a vivid idea on how they work.

KEYWORDS

- hidden Markov model
- named entity recognition
- natural language processing
- natural language toolkit
- verb finite
- word-sense disambiguation

REFERENCES

1. Jurafsky, D., (2000). *Speech and Language Processing*. Pearson Education India.
2. Voutilainen, A., (2003). "Part-of-speech tagging." *The Oxford Handbook of Computational Linguistics* (pp. 219–232).
3. Brill, E., (1992). "A simple rule-based part of speech tagger." *Proceedings of the Third Conference on Applied Natural Language Processing*. Association for Computational Linguistics.
4. Jelinek, F., (1985). Markov source modeling of text generation. Dordrecht. In: Skwirzinski, J., (ed.), *Impact of Processing Techniques on Communication*.

5. Bellegarda, J. R., (2014). *"Combined Statistical and Rule-Based Part-of-Speech Tagging for Text-to-Speech Synthesis."* U.S. Patent No. 8,719,00.

6. Cutting, D., et al., (1992). "A practical part-of-speech tagger." *Third Conference on Applied Natural Language Processing.*

7. Navigli, R., (2009). "Word sense disambiguation: A survey." *ACM Computing Surveys (CSUR), 41*(2), 10.

8. Rabiner, L. R., & Biing-Hwang, J., (1986). "An introduction to hidden Markov models." *Ieeeassp Magazine, 3*(1), 4–16.

9. Kupiec, J., (1992). "Robust part-of-speech tagging using a hidden Markov model." *Computer Speech and Language, 6*(3), 225–242.

10. Thede, S. M., & Mary, P. H., (1999). "A second-order hidden Markov model for part-of-speech tagging." *Proceedings of the 37th Annual Meeting of the Association for Computational Linguistics.*

11. Forney, G. D., (1973). "The Viterbi algorithm." *Proceedings of the IEEE, 61*(3), 268–278.

12. Miller, D. R. H., Tim, L., & Richard, M. S., (1999). *"A Hidden Markov Model Information Retrieval System* (Vol. 99)." SIGIR.

13. Nivre, J., (2000). "Sparse data and smoothing in statistical part-of-speech tagging." *Journal of Quantitative Linguistics, 7*(1), 1–17.

14. Nivre, Joakim, Johan Hall, & Jens Nilsson (2006). "Maltparser: A data-driven parser-generator for dependency parsing." *LREC, 6.*

15. Malhotra, S., and Godayal, D. (2018). An introduction to part-of-speech tagging and the Hidden Markov Model. Machine Learning. https://medium.freecodecamp. org/an-introduction-to-part-of-speech-tagging-and-the-hidden-markov-model-953d45338f24

16. Malhotra, S. and Godayal, D. (2018). A deep dive into part-of-speech tagging using the Viterbi algorithm. Machine Learning. https://medium.freecodecamp. org/a-deep-dive-into-part-of-speech-tagging-using-viterbi-algorithm-17c8de32e8bc

CHAPTER 7

Text Mining

S. KARTHIKEYAN,[1] JEEVANANDAM JOTHEESWARAN,[1]
B. BALAMURUGAN,[1] and JYOTIR MOY CHATTERJEE[2]

[1]*School of Computing Science and Engineering, Galgotias University, Greater Noida, Uttar Pradesh, India, E-mails: link2karthikcse@gmail. com (S. Karthikeyan), jeevanandamj@gmail.com (J. Jotheeswaran), kadavulai@gmail.com (B. Balamurugan)*

[2]*School of Computing Science and Engineering, LBEF (APUTI), Kathmandu, Nepal, E-mail: jyotirm4@gmail.com*

ABSTRACT

The major objectives of text mining (text data mining/text analytics) are to extract the pattern or information from the largely available unstructured or semi-structured text data. Data mining deals only with structured data whereas text mining deals with semi-structured or unstructured data, Around 80% of data stored throughout the globe is in unstructured or semi-structured form, it is the biggest need for text mining to manipulate the data in a meaningful way, there are many techniques like sentimental analysis, natural language processing (NLP), information extraction, information retrieval, clustering, concept linkage, associate rule mining (ARM), summarization, topic tracking are used to extract the data based upon the nature of data and will be discussed further on each technique in this chapter, but the major problem in the text mining is the ambiguity of the natural language, as the one word can be interpreted in multiple ways, ambiguity is the primary challenge for the researchers to address and the possible solutions are explained. Algorithms such as genetic algorithm, differential evolution can be combined to get the desired result, the output of algorithm can be scaled so that it can ensure the quality of the text retrieval. There are two methods called as precision and recall is used to measure text retrieval quality in text mining. There are several applications that are associated with text mining such as healthcare, telecommunication, research papers categorization, market analysis, Customer Relationship Management (CRM), banks, Information Technology and another environment where the huge unstructured volume of data is generated.

7.1 INTRODUCTION

Text mining (TM) is characterized as the non-minor extraction of covered up and possibly helpful information from textual data. TM is another field that endeavors to extract significant information from natural language text. It may be characterized as the process of dissecting text to extract information that is helpful for a particular reason. Comparing to the data in databases, the text is unstructured, unclear, and hard to process. In present-day culture, the text is the most mutual path for the formal trade of information [43].

TM typically manages texts whose work is the correspondence of genuine information or opinions, and the improvements for endeavoring to extract information from such text consequently are intriguing regardless of whether achievement is just fractional [8]. TM is like data mining (DM), then again, actually DM tools are intended to deal with structured data from databases, however, TM can likewise work with unstructured or semi-structured data sets, for example, messages, text records, and HTML documents and so on. Subsequently, TM is an obviously better arrangement [37].

TM ordinarily is the process of organizing the information text, inferring patterns inside the structured data, and the last assessment and elucidation of the output. The term TM is usually used to indicate any framework that breaks down huge amounts of natural language text and recognizes lexical or semantic utilization patterns trying to extract likely valuable information.

There are applications such as:

- Healthcare & biomedical;
- Social media;
- Banking;
- Customer management relationship;
- Education;
- Web-based software;
- Business intelligence;

- Sentiment analysis;
- Research paper classification;
- Security and biometric.

7.2 DATA MINING (DM) AND TEXT MINING (TM)

TM is the study and application of textual information extraction (IE) by the doctrine of computational linguistics. As an exploratory data analysis, TM is a method that uses the software to support decision-makers and researcher practitioners who uses large text collections in descending latest and pertinent information. The stakeholder is still involved, interacting with the system in the semi-automated process [55].

In the KDD (knowledge discovery from data) process, DM is a stride. KDD takes care of useful knowledge attainment that is novel, imperative, and legitimate [19]. DM needs little relations among the investigator and DM tool. It is an automatic process because DM tools inevitably search the data for anomaly and conceivable associations, thus identifying the unidentifiable problems by the end-users [37], while meager data analysis relies on the end-users to define the problem, choose the data, and commence the proper data analyses to help the model and resolve tribulations [20].

7.2.1 *ADVANTAGES AND DISADVANTAGES OF DATA MINING (DM)*

7.2.1.1 *ADVANTAGES OF DATA MINING (DM)*

7.2.1.1.1 *Marketing/Retail*

DM enables sales and marketing organizations to design models based upon previous data the target audience to the new advertising efforts, tools such as mail, online promoting effort and etc. Through a result, advertisers will have a fitting way to deal with pitching beneficial items to focused customers.

7.2.1.1.2 *Manufacturing Environment*

By applying DM in outfitted building data, makers can recognize defective gear or equipment and decide ideal control parameters. For instance, semiconductor makers have a test that even the states of manufacturing conditions at various wafer creation plants are comparative, the nature of wafer are a great deal the equivalent and some for obscure reasons even has abandons.

7.2.1.1.3   Banking and Financial Management

Financial reports on loan information and credit risks can be fetched using DM. Likewise, DM enables banks to identify untrustworthy credit card transactions to certify credit card's proprietor.

7.2.1.1.4   Social Media Analysis

Facebook and Twitter are considered as the most swarmed Social networking websites. These networking destinations have made it simple to speak with loved ones without endeavoring. Individuals identified with various qualities come nearer to one another by sharing their thoughts, premiums, and knowledge nowadays, it turns out to be simple for anybody to meet the general population of their interests for learning and sharing valuable information [56].

Additionally, social media is a fusion of few learning frames, for instance, e-learning, and m-learning. On various social networking destinations, the most widely recognized strategy for association with one another is through text. People can exchange their ideas by blogs, posts, and all other medium where text is involved. The utilization of the TM techniques is to produce the text correct so that it is useful for anybody to compose in the most proper way TM implies the extraction of the data which isn't natural to anybody.

7.2.1.2   DISADVANTAGES OF DATA MINING (DM)

7.2.1.2.1   Information Usage/Inaccurate Information

There are intruders in the globe who will access the data of the autho-rized individuals or organizations where they will use the unauthorized information without the knowledge of the genuine user. When the users using genuine information, decisions can be taken effectively, whereas the results will be ineffective if incorrect information being used for taking decisions.

7.2.1.2.2   Privacy in Database

The users trust on the internet is reducing constantly as being the data can intruded by the attackers at any point of time during the message transmission among blogs, forums, online business, social networks and etc. Due to this reasons, authenticated users still worried about the data security whether it is stored securely or not. Organizations collect data on customers from various means for knowing similar patterns.

7.2.1.2.3   Security Issues

Organizations possess information on their representatives and customers including Citizen Identification number, birthday, and other financial data. The data that are stored about the people are still in question regarding the data security and other means. There are events occurring regularly in the globe where the intruders attacking the central bank server or any server which has the sensitive data. This leads to huge financial loss for organization, making people to lose the trust on the organization [57].

7.2.2   DATA MINING (DM) VERSUS TEXT MINING (TM)

Earlier years, IT individuals concentrated on DM, where they will extract the knowledge from the huge volume of text which is unstructured, as most of the organization does not possess the text in a structured way, it challenges them to get the desired result. Much organization takes decision-based upon the available information from the large database.

The major difference between TM and DM is that the TM gives knowledge from both structured as well as unstructured text. It can deal with all three different types such as structured, unstructured, and semi-structured data.

7.2.2.1 TEXT MINING (TM) APPROACHES

Similarly, as DM isn't only a novel methodology or a solitary system for finding knowledge from data, TM additionally comprises of a wide assortment of techniques and innovations [47], All TM methodologies have a typical feature such as dealing about processing text. For example:

- **Keyword Based Advances:** The information depends on a choice of keywords in text that are separated as a progression of character strings.
- **Statistics Innovations:** Alludes to frameworks dependent on AI. Statistics advancements influence a preparation set of archives utilized as a model to oversee and classify text.
- **Linguistic Based Innovations:** It makes use of language processing frameworks. The output of text analysis permits a shallow comprehension of the logic, grammar, text structure.

7.3   PRELIMINARY TEXT MINING (TM) METHODS

7.3.1   TEXT PREPROCESSING

Text preprocessing is a basic component of any NLP framework, since the characters, words, and sentences recognized at this stage are the key units gone to all further processing stages, from analysis and labeling segments, like morphological analyzers and grammatical form taggers, through applications, like information recovery and machine interpretation frameworks [6]. It is an accumulation of exercises in which text archives are preprocessed. Since the text data frequently contains some extraordinary configurations like number arrangements, date groups and the most well-known words that farfetched to help TM, like, relational words, articles, and professional things can be wiped out [6].

- **Purpose of Text Preprocessing in NLP System:**
  1. To diminish the indexing record size of the text archives:
     - Stop words produce 20–30% of the complete word includes in a specific text archives
     - Stemming may decrease ordering size maybe 50%.
  2. To improve the productivity and viability of the IR framework

- Stop words didn't produce that much impact on TM and it affects the retrieval system.
- Stemming utilized for coordinating the comparable words in a text report.

7.3.2 TOKENIZATION

Tokenization is the process of separating a character succession into pieces of words or phrases called tokens, and maybe in the meantime discard certain characters. Here the text is divided into words, expressions, images, or other significant components called tokens. The point of the tokenization is the investigation of the words in a sentence [1].

Tokenization is helpful both in linguistics that is a type of text division, and in software engineering, where it frames some portion of lexical analysis. Textual data is just a square of characters toward the start. All processes in information recovery require the words of the data set. Henceforth, the necessity for a parser is a tokenization of records [1] (Figure 7.1).

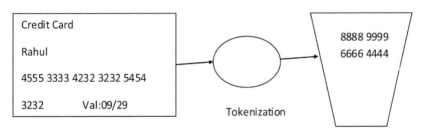

FIGURE 7.1   Tokenization.

This may sound inconsequential as the text is as of now put away in machine-meaningful arrangements. Even few issues are still left, similar to the expulsion of accentuation marks. Different characters like sections, hyphens, and so on require processing as well. Besides, tokenizer can provide food for consistency in the archives. The primary utilization of tokenization is recognizing the important keywords. The irregularity can be diverse number and time positions. Another issue is shortenings and abbreviations which must be changed into a standard structure.

7.3.3   STOP WORD REMOVING

In a document, large number of words repeats around regularly yet is basically negligible as they are just used to consolidate words in a sentence. It is generally known that stop words don't add to the context or substance of textual records. As the words are repeated multiple number of times, it challenges the TM to further processing of document.

Statement with Stop Words	Statement without Stop Words
Akilan was studying Computer Science	Akilan, studying, Computer science
Researchers are working hard on new innovations	Researchers, working, hard, new innovations
Mining is the best method for taking decisions	Mining, method, taking, decisions
Technologies are making people dumb	Technologies, making, people, dumb

Stop words use the most regular words like 'is,' 'was,' 'and,' 'are,' 'this' and so on. They are not helpful in arrangement of words and should be evacuated. In the above example you can see the statement "Akilan was studying Computer Science" where 'was' is the stop which is not needed in the text process, on next statement "Researchers are working hard on new innovations," where 'are' and 'on' belongs to stop words, which can be eliminated in the TM process [1].

7.3.4   STEMMING

It comes under natural language processing (NLP) and natural language understanding (NLU) where the word or the term is reduced to its stem such as run, running, and runner are the words which form the stem such as run as a whole.

There are two errors in stemming:

1. Over stemming; and
2. under stemming
   - **Over Stemming:** It occurs if the two words with various stems are stemmed to a similar root. This is otherwise called a false positive.

- **Under Stemming:** It occurs if the two words are stemmed to a same root are most certainly not. This is otherwise called a false negative.

7.3.4.1 TYPES OF STEMMING ALGORITHMS

1. **Table Lookup Approach:** One strategy to do stemming is to store a table of all record terms and their stems. Terms from the inquiries and records could then be stemmed through query table, utilizing b-trees or hash tables. Such queries are extremely quick; however, there are issues with this methodology [1]. First, there is no such data for English, regardless of whether there were they may not be spoken to on the grounds that they are space explicit and require some other stemming strategies. Second issue is capacity overhead.

2. **Successor Assortment:** These stemmers depend on the auxiliary linguistics which decides the word and morpheme limits dependent on dispersion of phonemes. Successor assortment of a string is the quantity of characters that tail it in words in some group of text. For instance, consider a group of text comprising of the following words [1].

Karthik, Kavin, Kenny, Karina, Krunal, Jeeva, Bala, Joy

How about we decide the successor assortment for the word read. The first letter in Karthik is K. K is followed in the text body by 3 characters A, E, R subsequently the successor assortment of R is 3. The following successor assortment for KAR is 2 since T, I pursues KAR in the text body, etc. Table 7.1 demonstrates the total successor assortment for the word read.

TABLE 7.1 Stemming Approach

Prefix	Successor Variety	Letters
K	3	A, E, R
KA	2	R, V
KAR	2	T, I
KART	1	H

When the successor assortment for a given word is resolved then this information is utilized to portion the word.

i. **Cut Off Strategy:** Some cutoff esteem is chosen and a limit is recognized at whatever point the cut off esteem is come to.

ii. **Peak and Plateau Strategy:** In this technique, a section break is made after a character whose successor assortment surpasses that of the characters promptly going before and tailing it.

iii. **Complete Word Strategy:** Break is made after a fragment if a section is a finished word in the corpus.

3. **N-Gram stemmers:** This strategy has been planned by Adamson and Boreham. It is called a shared diagram strategy. Digram is a couple of continuous letters. This technique is called n-gram strategy since trigram or n-grams could be utilized. In this technique, affiliation measures are determined between the sets of terms dependent on shared one of a kind diagram [1].

For instance: consider two words Stemming and Stemmer;

Running->ru un nn ni in ng

Runner-> ru un nn er

In this precedent the word running has 6 interesting diagrams, runner has 4 kind diagrams, these two words share 3 novel diagrams ru un nn. When the quantity of exceptional diagrams is discovered then a similitude measure dependent on the novel digrams is determined utilizing dice coefficient. Shakers coefficient is characterized as:

$$S = 2C/(A+B)$$

where: C = Regular one of a kind diagrams; A = Quantity of special diagrams in the first word; B = Quantity of one of a kind diagrams in the second word.

Comparability measures are resolved for all sets of terms in the database, framing a closeness matrix. When such a comparability matrix is accessible, the terms are clustered utilizing a solitary connection clustering technique.

4. **Affix Removal Stemmers:** Fasten evacuation stemmers expel the additions or prefixes from the terms leaving the stem. One of the cases of the append expulsion stemmer is one which expels the plurals type of the terms. Some arrangement of tenets for such a stemmer is as per the following:

    i. If any word that finishes in "ies" yet not "eies" or "aies "
        At that point "ies" - > "y"
   ii. If any word that finishes in "es" however not "aes," or "ees" or "oes"
        At that point "es" - > "e"
  iii. If a word finishes in "s" however not "us" or "ss "
        At that point "s" - > "Invalid"

7.4   TEXT MINING (TM) TECHNIQUES

7.4.1   *CLUSTERING*

Clustering is an unsupervised process to order the text files in gatherings by utilizing distinctive clustering algorithms. There are two approaches that are used in clustering, such as top-down and bottom-up approaches. In NLP, numerous kinds of mining methods are utilized for the assurance on the unstructured text [4].

7.4.1.1   *TEXT CLUSTERING*

The Clustering method is an optimal methodology that can be used for processing larger volumes of data [32]. It is found that text clustering is a standout amongst the best methodology utilized for text analysis [9].

The process of topic tracking has picked up the enthusiasm from the scientists who are taking a shot at the subject of text clustering in the computerized field [45]. Different techniques and algorithms dependent on unsupervised records the executives are incorporated into the process of report clustering. In this process, the numbers, properties, and relationship of the assembled sets are at first obscure.

7.4.2   *K-MEANS ALGORITHM*

The K-implies algorithm divides a gathering of vectors {V1, V2, Vn} into the set of clusters {CL1, CL2, CLk}. The algorithm needs k cluster seeds for the introduction. They can be remotely provided or grabbed arbitrarily among the vectors. The algorithm continues as pursues Introduction K seeds, either given or chose arbitrarily; structure the center of k clusters. Each and every other vector is appointed to the cluster of the nearest seed. Emphasis The centroid Mi of the present cluster is figured every vector is reassigned to the cluster with the nearest centroid. Stopping conditions at convergence-when no more changes happen. The K-implies algorithm augments the clustering quality capacity Q, if the separation metric (opposite of the likeness work) acts well regarding the centroids calculation, at that point every emphasis of the algorithm expands the estimation of Q [2].

An adequate condition is that the centroid of a set of vectors be the vector that amplifies the total of likenesses to every one of the vectors in the set. This condition is valid for every "natural" metric. It pursues that the K-implies algorithm dependably merges to a local maximum. The K-implies algorithm is famous in light of its effortlessness and proficiency. The complexity of every emphasis is O(kn) likeness correlations, and the quantity of vital emphases is normally very little.

7.4.3   *HIERARCHICAL AGGLOMERATIVE CLUSTERING (HAC)*

The HAC algorithm starts its work with each item specifically cluster and continues, as per some picked foundation it is over and again combine sets of clusters that are generally comparable. The HAC algorithm completes when everything is converged into a solitary cluster. Parallel tree of the clusters chain of command is given by history of consolidating. The algorithm continues as pursues: Instatement Every single article is put into a different cluster. Emphasis Discover the pair of most comparable clusters and union them.

Stopping condition rehashes stage 2 until a solitary cluster is framed. When everything is converged into single cluster distinctive renditions of the algorithm can be delivered, at that, point it is determined the closeness between clusters. The complexity of this algorithm is O(n2s), where n is the quantity of articles and s the complexity of computing similitude between clusters.

Estimating the Nature of an algorithm needs human judgment, which presents a high level of subjectivity. Given a set of ordered (physically grouped) records, it is conceivable to utilize this benchmark marking for assessment of clustering's. The most widely recognized measure is virtue. Accept {L1, L2, Ln} are the physically marked classes of reports, and {C1, C2, Cm} are the clusters returned by the clustering process [2].

7.4.4 CATEGORIZATION/CLASSIFICATION

The objective of text categorization is to arrange a set of archives into a fixed number of predefined classifications, where each report may have a place with more than one class. The categorization undertaking is to order a given data example into a pre indicated set of classifications [12]. Text categorization is a sort of "managed" realizing where the classifications are known ahead of time and firm in advancement for each preparation archive. It includes distinguishing the principle subjects of the report into a pre-characterized set of themes. While sorting a record, a PC program will regularly regard the report as a "Pack of Words" [13].

Categorization is the task of ordinary language archives to a predefined set of points as per their substance [11]. It is an accumulation of text reports, the process of finding the exact point or themes for each archive [45]. These days automated text categorization is connected in a variety of contexts from the established programmed or semiautomatic indexing of texts to customized ads conveyance, spam filtering, and categorization of Website page under hierarchical lists, programmed metadata age. Categorization devices have a technique for positioning the archive arranged by which reports have the most substance on a specific point [2] (Figure 7.2).

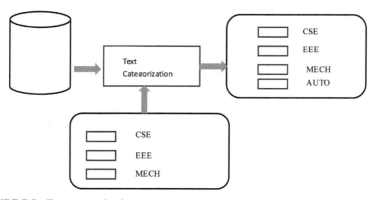

FIGURE 7.2 Text categorization.

7.4.4.1 CLASSIFICATION

Text classification has been extensively considered in various networks, like DM, database, machine learning and information retrieval (IR), also utilized in a tremendous number of uses in different areas, for example, picture processing, therapeutic analysis, record association, and so forth. Text classification expects to relegate predefined classes to text reports. The issue of classification is characterized as pursues. A preparation set D = {d1, d2, dn} of reports, to such an extent that each record di is named with a mark ℓi from the set $L = \{\ell 1, \ell 2, \ell k\}$. The undertaking is to discover a classification show (classifier) f where:

$$f: D \longrightarrow L\ f\,(d) = \ell$$

$f: D \longrightarrow L\ f\,(d) = \ell$ which can appoint the right class mark to new report d. The classification is called hard if a mark is expressly appointed to the test occasion and delicate, if an esteem is allocated to the test occurrence. There are different kinds of classification which permit task of numerous marks to a test occasion. For a broad diagram of various classification strategies. Huge numbers of the classification algorithms have been actualized in various programming frameworks and are openly accessible, to assess the execution of the classification show, we set aside an irregular portion of the marked records (test set). In the wake of preparing the classifier with the preparation set, we classify the test set and contrast the assessed marks and the genuine names and measure the execution [1].

7.4.5   *ASSOCIATION RULE MINING (ARM)*

ARM is a strategy to find the visited patterns, associations, fundamental structures from data sets found in different sorts of databases, like transactional databases, social databases, and different types of data storehouses [21]. The ARM distinguishes the variable-esteem blends which will, in general, happen much of the time. The strategy for ARM otherwise called knowledge discovery in databases; that is like the correlation analysis that discovers the connections between two factors [9]. In the knowledge models DM found, association rule demonstrate is a vital one, additionally the most dynamic branch. Association rule demonstrates the correlation between gatherings of items in the database.

Association rule mining (ARM) is to find the fascinating or significant association between substantial quantities of data things, numerous scientists had top to bottom research on it, at that point improved, and extended the underlying association rules mining algorithm. In the interim, the

ARM has been connected to the database in numerous different fields and accomplished great uncovering impact.

ARM is to distinguish the association rules in the transaction database D to meet client indicated least help and least certainty; the whole mining process can be decayed into two stages, explicitly appeared in Figure 7.3.

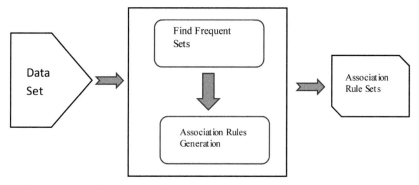

FIGURE 7.3 Association rule mining steps.

In the first place, to discover the item set with all the transaction underpins more prominent than the base help. The help of a thing set is the quantity of transaction containing this set. The item set with least help called visit itemsets, others are non-visit item sets. That is, to recognize the thing set with help more prominent than a given help threshold. The second step, create solid association rules dependent on regular itemsets finding. That is, create the association rules with help and certainly more prominent than or equivalent to the earlier given threshold.

7.4.6 TOPIC TRACKING

A topic tracking framework process by supporting client details and, in light of the documents the user identifies, predicts different documents crucial to the client. There are numerous zones where topic tracking can be connected in industry. It tends to be utilized to alert organizations whenever a contender is in the news. This enables them to stay aware of focused items or changes in the market. Likewise, organizations should need to follow news all alone company and items. It could be utilized in the medicinal business by specialists and searching for new medications for sicknesses and who wish to keep up on the most recent headways. People in the field of training can utilize topic tracking to make sure they have the most recent references for research in their general vicinity of intrigue [3].

Keywords are a deposit of noteworthy words in an editorial that gives an abnormal state depiction of its substance to perusers. Distinguishing keywords from a lot of online news data is exceptionally helpful in that it can create a short outline of news articles. Manual keyword extraction is a very troublesome and tedious errand; truth be told, it is practically difficult to extract keywords physically if there should be an occurrence of news articles distributed in a solitary day because of their volume.

News pages, which use HTML, are taken from an Internet gateway website. Furthermore, the respective module pulls out hopeful keywords. Lastly, keywords are pulled out by cross-area examination module. The keyword extraction module is portrayed in specific. The tables for 'document,' 'lexicon,' 'term happen certainty' and 'TFIDF weight' in the relational database are made [3].

7.4.7   CONCEPT LINKAGE

It associates correlated documents by recognizing their usually common concepts and makes clients discover information that they maybe wouldn't have discovered utilizing customary searching techniques. It advances perusing for information instead of searching for it. Concept linkage is a significant concept in TM, particularly in the biomedical area where huge research happens that it is unimaginable for researchers to peruse all the material and do association to other research [3].

In a perfect world, concept connecting programming can recognize interfaces among illnesses and medicines when people can't [3].

7.4.8   INFORMATION EXTRACTION (IE)

The primary objective of IE techniques is the extraction of helpful information from text [22]. It recognizes the extraction of occasions, elements, and connections from semi-structured or unstructured text. IE programming distinguishes key expressions and connections inside the text. IE issues the extraction of semantic information from the text [2] (Figure 7.4).

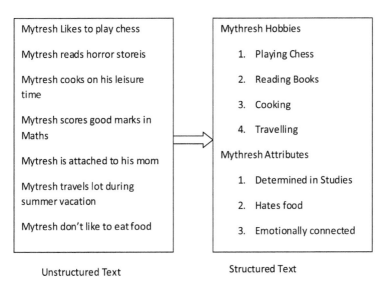

Mytresh Likes to play chess	Mythresh Hobbies
Mytresh reads horror storeis	1. Playing Chess
Mytresh cooks on his leisure time	2. Reading Books
	3. Cooking
Mytresh scores good marks in Maths	4. Travelling
Mytresh is attached to his mom	Mythresh Attributes
Mytresh travels lot during summer vacation	1. Determined in Studies
	2. Hates food
Mytresh don't like to eat food	3. Emotionally connected
Unstructured Text	Structured Text

FIGURE 7.4   Information extraction.

A commencement point for unstructured manuscripts to assess unstructured original copies is to utilize IE. IE programming perceives key expressions and connections incorporated into the original copy. It is done by finding the predefined courses of action in a text; this system is called pattern coordinating. Ordinary language text documents comprise of information that can't be used for mining. IE concurs with the documentation, picking proper articles, and the association among them to make them increasingly accessible for included direction [8].

In spite of IR, it achieves to perceive important documents from a document accumulation, IE results ordered information arranged for post-processing, which is basic to different uses of Web mining and probing instruments. It comprises of isolating suitable text parts, extracting the offered data in such parts, and changing the data into the functional structure. Fractional extraction from space-specific texts is at present conceivable; however, total IE from the arbitrary text is as yet a proceeding with study target [46].

7.4.9   INFORMATION RETRIEVAL (IR)

IR does the action of discovering information assets from a gathering of unstructured data sets that fulfill the information need [18] (Figure 7.5).

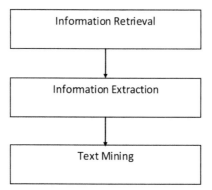

FIGURE 7.5   Information retrieval.

In this way, IR, for the most part, centered on encouraging information to get to as opposed to examining information and finding concealed patterns, which is the primary reason for TM. IR has less need on processing or change of text while TM can be considered as going past information access to additionally help clients to break down and comprehend information and facilitate the decision making [1].

The IR framework is a network of algorithms, which encourage the search of significant data/documents according to the client prerequisite [36]. It not just gives the significant information to the client yet additionally tracks the utility of the showed data according to client conduct, for example, is the client-finding the outcomes valuable or not. The most outstanding IR frameworks are Google search engines which perceive those texts on the WWW (World Wide Web) that are related with available words [2].

It is estimated as an augmentation to document retrieval where the documents that are returned are processed to extract the valuable information pivotal for the client. Subsequently, document retrieval is trailed by a text summarization arrange that centers around the question presented by the client, or an IE organize. IR in the more extensive sense manages the entire scope of information processing, from IR to knowledge retrieval. It increased expanded consideration with the development of the WWW and the search engine.

IR is assigned as the full structure to document retrieval where the documents are returned and processed to gather or get the specific information recovered by the client. In this manner document retrieval could be trailed by a text summarization arrange that centers around the question presented by the client, or an IE organize utilizing strategies. IR frameworks help in to limit the set of documents that are pertinent to a specific issue [4].

7.4.10 INFORMATION VISUALIZATION

Information visualization puts substantial textual sources in a chain of importance or maps [16]. The Information given by graphical visualization is better, far-reaching and quicker reasonable than unadulterated text-based portrayal so it is best to mine the huge document accumulation. Information visualization is valuable when a client needs to limit an expansive scope of documents and investigate related topics. The vast majority of the methodologies of TM are roused by the strategies which had been proposed in the zone of visual DM, information visualizations, and explorative DM [2].

This strategy can improve the discovery or extraction of applicable patterns or information for TM and IR frameworks. Information that permits a visual portrayal involves parts of the outcome set, keyword relations or metaphysics is viewed as the parts of the search process itself. The objective of information visualization, the development might be directed into three stages:

1. Preparation of data;
2. Analysis and extraction of data; and
3. Mapping the data visualization.

Information visualization puts incredible textual bases in a visual progression or plan and offers perusing capacities just as general searching. This strategy offers improved and snappier extensive knowledge, which helps us to mine gigantic gathering documents. The administrators can recognize the hues, associations, and holes. The combination of documents can be shown as a structured design using the indexing or vector space demonstrate (VSM) [9] (Figure 7.6).

7.4.11 SUMMARIZATION

Text Summarization: Numerous TM applications need to outline the text documents so as to get a succinct review of a substantial document or a gathering of documents on a topic. There are two classes of summarization procedures by and large: extractive summarization where a synopsis involves information units extracted from the first text, and in opposite abstractive summarization where a rundown may contain "orchestrated" information that may not happen in the first document [1].

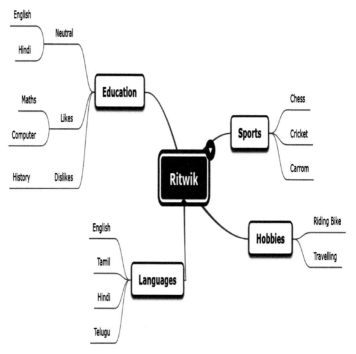

FIGURE 7.6   Information visualization.

Text summarization helps to reduce the length and aspect of a document while holding its primary concerns and in general significance [10]. Text summarization is the process of naturally making a compacted variant of a given text that gives helpful information to the client [2]. A programmed summarization process can be partitioned into three stages

Text summarization includes different strategies that utilize text categorization, for example, NN (neural networks), DT (decision trees), SG (semantic graphs), Relapse models, swarm insight and fuzzy. These techniques have a typical issue, that is, the nature of the improvement of classifiers is variable and exceptionally subject to the sort of text being abridged (Figure 7.7).

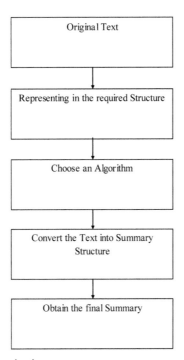

FIGURE 7.7 Text summarization.

Text summarization categorized into two types such as:

1. **Extractive Summarization:** These strategies depend on extracting a few sections, for example, expressions, and sentences, from available text and put them together to make a summary. In this manner, recognizing the correct sentences for summarization is of most extreme significance in an extractive technique.

2. **Abstractive Summarization:** These strategies utilize propelled NLP [42] procedures to produce an altogether new synopsis. A few pieces of this rundown may not by any mean show up in the first text.

7.4.12 PAGERANK ALGORITHM

Here four web pages are associated—web 1, web 2, web 3, and web 4. The web pages have connections redirecting to one another. Few web pages does not have no link, also known as dangling pages [52] (Figure 7.8 and Table 7.2).

- Web 1 has links going to Web 3, Web 2.
- Web 2 has links for Web 1, Web 3.
- Web 3 connected only to Web 2.
- Web 4 has no links so it is called a dangling page.

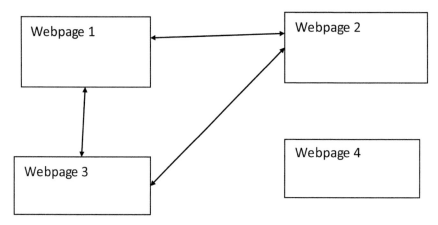

FIGURE 7.8   Web pages links.

TABLE 7.2   Web Pages Connectivity

Webpages	Links
Web 1	Web 3, Web 2
Web 2	Web 1, Web 3
Web 3	Web 2
Web 4	Nil

The probability of a user visiting that page called a Page Rank score is calculated. For the apprehension of the probabilities of users navigate from one web page to other web pages, then it produces a square matrix A that have 'n' rows and 'n 'columns, that results in K*K, where **K** is the number of web pages [17].

The possibility of one individual transitioning from one web page to other web page is denoted as each element of this matrix [51].

7.4.12.1 TEXT RANK ALGORITHM

The sentence comparison between among two sentences is equivalence to the probability of web page redirecting. The correspondence scores will be placed inside square matrix, as same as the matrix A used for PageRank. It is an extractive and unsupervised text summarization technique. The flow of the TextRank algorithm is as follows (Figure 7.9):

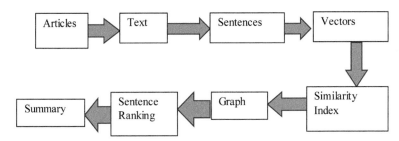

FIGURE 7.9 TextRank algorithm.

- **Step 1:** The concatenation of text contained that are present in the articles.
- **Step 2:** Divide the available text into individual sentences.
- **Step 3:** Word embedding for each and every sentence is found as vector representation.
- **Step 4:** Resemblances between sentence vectors must be computed then stored as a matrix.
- **Step 5:** The similarity matrix is to change as a graph.
- **Step 6:** The number of high ranked sentences are derived produce the summary.

7.4.13 NATURAL LANGUAGE PROCESSING (NLP)

NLP can be called a category of artificial intelligence (AI), and linguistics which goes for a comprehension of natural language utilizing computers. A large number of the TM algorithms broadly make utilization of NLP procedures, for example, part of speech labeling (POG), syntactic parsing, and different sorts of linguistic analysis. NLP challenges the researchers in the field of AI [40]. The objective is to make understand the computers as humans do with the NLP [1].

NLP research seeks after the obscure inquiry of how we comprehend the importance of a sentence or a document. What are the signs we use to comprehend who did what to whom [8], or when something occurred, or what is reality and what is supposition or expectation? While words things, action words, intensifiers, and descriptors [8] are the building squares of importance, it is their correlation to one another inside the structure of a sentence in a document, and inside the context of what we definitely think about the world, that gives the genuine significance of a text [4] (Figure 7.10).

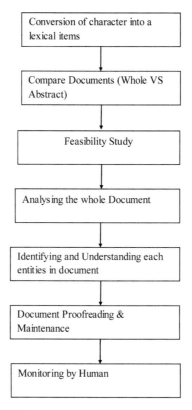

FIGURE 7.10   Steps in NLP.

There are basically seven steps that are carried out in NLP such as conversion the character into lexical string, comparing the documents (Macro with Micro level Documents), performing feasibility study whether the documents will be able to extract the information, followed by the analyzing of documents, then every document in the database will be identified as individual entity, after the identification, proofreading can be performed, where all the steps are carried and monitored by the humans.

7.5 TEXT MINING (TM) ADVANCED ALGORITHMS

The TM algorithms are classified as (Figure 7.11):

- Classification algorithm;
- Discovering associations;
- Clustering algorithm; and
- Differential evolution.

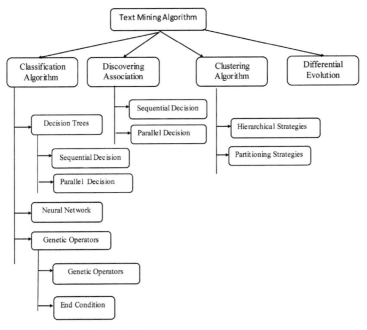

FIGURE 7.11 Text mining algorithm types.

7.5.1 CLASSIFICATION ALGORITHM

The classification issue can be communicated as a planning data set containing records. Each record is perceived by a fascinating record id, and includes fields contrasting with the characteristics. A quality with a constant zone is known as a tenacious quality. An attribute with a constrained space of discrete characteristics is known as a decided trademark. One of the decided qualities is the ordering trademark or class and the motivation in its space are called class names.

7.5.2 CLASSIFICATION MODELS

The unmistakable kind of classification models, for instance, DT, NN, and genetic algorithm (GA) and so on.

7.5.2.1 CLASSIFICATION UTILIZING DECISION TREE (DT)

7.5.2.1.1 Sequential Decision Tree (DT) Based Classification

A DT includes an inside hub and leaves. All of the inward hubs have a decision related with it and all of the leaves has a classmark associated with it.

A DT based classification includes two phases:

1. **Tree acceptance:** A tree is prompted from the given planning set; and
2. **Tree pruning:** The started tree is made logically concise and vivacious by clearing any verifiable conditions on the specific getting ready data set.

Parallel arrangement of DT based classification the goal of parallel itemizing of DT based classification algorithms are flexibility in both runtime and memory essentials [5, 27].

Diverse kinds of parallel designs for the classification DT advancement are:

1. Synchronous tree improvement approach;
2. Partitioned tree improvement approach; and
3. Hybrid parallel itemizing.

*   **Synchronous Tree Advancement Approach:** In this system, all processors fabricate a DT synchronously by sending and accepting class appropriation data of nearby data.

- **Divided Tree Improvement Approach:** In this approach, at whatever point conceivable, particular processors tackle different pieces of the classification tree.
- **Classification Using Neural Network:** In supervised learning, a lot of point of reference sets (x,y), x $\in$ X, y $\in$ Y and the fact is to find a capacity f in the allowed class of capacities that arranges the models. In that capacity, we wish to infer the mapping proposed by the data. The cost capacity is related to the perplexity between our mapping and the data and it irrefutably contains prior information about the issue zone [5].
- **Classification using Genetic Algorithm (GA):** GA is a heuristic optimization strategy whose frameworks are for all intents and purposes equal to organic advancement. In GA, the courses of action are called individuals or chromosomes. After the hidden masses are made discretionarily, selection, and assortment work is executed around until some end standard is come to. Each continues running of the circle is known as an age [5].

7.5.2.2 GENETIC ADMINISTRATORS

The GA uses crossover and mutation administrators to deliver the offspring of the present populace. Before genetic administrators are associated, watchmen have been decided for evolution to the general population to come. The crossover and mutation algorithm is used to make the general population to come [5].

7.5.2.3 END CONDITION

If there is no satisfactory improvement in somewhere around two sequential ages, stop the GA procedure. In various cases, time repression can be used when in doubt for finish the procedure.

7.5.3 ALGORITHM FOR FINDING ASSOCIATIONS

The goal is to check whether the occasion of explicit things in an exchange can be used to diminish the occasion of various things, or by the day's end, to find partnered associations between things [5]. This kind of data is especially beneficial for an Internet server supporting a web business site to associate the different thing pages dynamically, in perspective on the client collaboration.

7.5.3.1   PARALLEL ALGORITHM FOR FINDING

Associations: The issue can be communicated as given a lot of things, association rules predict the occasion of some other arrangement of things with certain dimension of sureness.

7.5.3.2   SEQUENTIAL ALGORITHM FOR FINDING

Association: The idea of association guidelines can be summed up and made progressively accommodating by viewing another reality about exchanges. All exchanges have a timestamp related with them; for instance, the time at which the exchange occurred. In case this data can be put to use, one can find associations, for instance, if a client acquired book today, by then he/she is most likely going to buy a book in two or three days' time [5].

7.5.4   CLUSTERING ALGORITHM

7.5.4.1   CLUSTERING

Clustering is a division of data into similar objects, called cluster, contains articles that are similar among themselves and not under any condition like objects of various gatherings [14]. Addressing data by less clusters on a very basic level loses certain fine nuances (much equivalent to lossy data pressure), [15] anyway achieves adjustments [5].

7.5.5   DIFFERENTIAL EVOLUTION

It is used to optimize or find a solution for a complex problem, where the problems undergo frequent iteration so that an optimal solution can be obtained with required quality as per the need. There are three essential operators associated with the process of the DE algorithm, including the mutation operator, the crossover operator, and the selection operator. DE improves a problem by keeping up a population of arrangements and making new arrangements by joining existing ones as indicated by its basic formulae and after that keeping whichever arrangement has the best score or wellness on the optimization problem within reach.

The process of basic DE is outlined as follows (Figure 7.12):

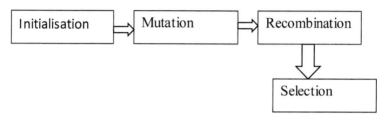

FIGURE 7.12   Differential evolution.

Along these lines the optimization problem is treated as a black box that only provides a proportion of value given a competitor arrangement and the angle is in this way not needed. These operators have a few similitudes to the GA, in any case, the DE contrasts essentially with the GA in the mutation process, in that the freak arrangement is created by including the weighted distinction between two irregular population individuals to the third part [5].

7.6   TEXT TRANSFORMATION

Text transformation converts the available data to the meaningful form which is used to predict some decision for business or any other means, irrespective of the techniques like sentimental analysis, clustering, classification, categorization, topic tracking, association rules mining, the role of text transformation is essential in the process of TM. It can be known as a dimension reduction technique.

7.6.1   BAG OF WORDS

A bag of words show is a method for extracting features from the text for use in displaying, for example, with machine learning algorithms (LAs).

The methodology is straightforward and adaptable and can be utilized in a bunch of ways for extracting features from documents. A bag of-word is a portrayal of text that depicts the event of words inside a document.

It includes two entities:

- Known texts; and
- A proportion of scaling of known words.

The model is just worried about whether realized words happen in the document, not wherein the document [4].

Following are the guide to comprehend this concept top to bottom:

"It is the best of places"

"It is the worst of places"

Here each are separate document and treated individually.

'It,' 'is,' 'the,' 'best,' 'of,' 'places,' 'worst'

Then create vectors that are used to convert text utilized by the machine LA.

Consider the statement "It was the best of places" find the occurrences of words from the seven unique words.

- "it" = 100
- "is" = 100
- "the" = 100
- "best" = 100
- "of" = 100
- "Places" = 100
- "worst" = 0

Rest of the documents will be:

It is the best of places = [100, 100, 100, 100, 100, 100, 0, 0, 0, 0]

It is the worst of places = [100, 100, 100, 0, 100, 100, 100, 0, 0, 0]

In this approach, every word or token called as "gram." Constructing a term of 2-word pairs is bigram model.

In the first statement, bigram can be explained as: "It was the best of places":

- it is
- is the
- the best
- best of
- Of places

The change happening over NLP text to numbers is vectorization in ML. Distinctive approaches to change over text into vectors are:

Tallying the occasions each term shows up in a document.

Ascertaining the occurrence that every term displays up in a document out of the considerable number of terms in document.

7.6.2 TF-IDF VECTORIZER

TF-IDF represents term frequency (TF)-inverse document frequency. TF-IDF weight is a factual measure used to assess how vital a word is to a document in an accumulation or corpus [4].

7.6.2.1 TERM FREQUENCY (TF)

It is a scoring of the occurrence of the term in the present document. As each document is diverse long, it is conceivable that a word would look substantially more occasions in long documents than smaller ones. The TF is often isolated by the document length to standardize.

TF (t) = Number of times 't' appear in a document

Total number of terms in the document

7.6.2.2 INVERSE DOCUMENT FREQUENCY

It is achieving of a number of times the text in the documents. If the terms are less, higher is the IDF score.

$IDF(t) = \log_e$ (Available Documents)

Available Documents with 't'

Thus,

TF-IDF score = TF*IDF

1. **Vector Space:** Salton had developed this model and it used the universal and the local text on words in available documents. Text documents are represented by an algebraic model with vectors of identifiers. The VSM is a method used to speak to documents and inquiries as vectors in multidimensional space, whose measurements are the terms used to fabricate a list to speak to the

documents. It is the most widely utilized method for IR because of its straightforwardness; proficiency over substantial document accumulations and it is speaking to utilize. The adequacy of the VSM depends for the most part on the term weighting connected to the term of the document vectors [4].

The VSM has three stages (Figure 7.13):

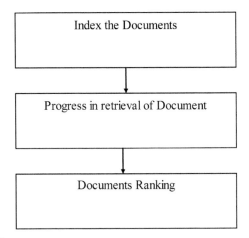

FIGURE 7.13   Vector space stages.

The above three steps explain the hierarchy and the process of vector space, where initially indexing has to be done, then finding the improvement in the document retrieval, which leads to the final phase of documents ranking.

7.7   TEXT RETRIEVAL MEASUREMENT

Text retrieval measurements can be done by the following parameters such as true Positive, True Negative, False Positive, False Negative, Accuracy, Precision, Recall, F1 Score.

7.7.1   TRUE POSITIVES (TP)

TP are the effectively anticipated positive qualities which indicate that the estimation of result is yes and the estimation of anticipated result is additionally yes. For example in the event that genuine class esteem shows that this traveler endure and anticipated class reveals to you a similar thing.

7.7.2 TRUE NEGATIVES (TN)

TN are the effectively anticipated negative qualities which indicate that the estimation of result is no and estimation of anticipated result is likewise negative.

For example, in the event that real result says this traveler did not endure and anticipated class discloses to you a similar thing. False positives (FP) and false negatives (FN), these qualities happen when your result repudiates with the anticipated class.

7.7.3 FALSE POSITIVES (FP)

When real class is no and anticipated result is positive. For example, in the event that genuine class says this traveler did not endure but rather anticipated class discloses to you that this traveler will endure.

7.7.4 FALSE NEGATIVES (FN)

When genuine class is yes yet anticipated result in negative. For example on the off chance that real class esteem shows that this traveler endure and anticipated class discloses to you that traveler will kick the bucket

7.7.5 ACCURACY

It is the most natural execution measure and it is essentially a proportion of accurately anticipated perception to the complete perceptions.

$$\text{Accuracy} = TP+TN/TP+FP+FN+TN$$

7.7.6 PRECISION

Precision is the proportion of effectively anticipated positive perceptions to the complete anticipated positive perceptions. The inquiry that this measurement answer is of all travelers that marked as endured, what number of really endure? High precision identifies with the low false positive rate.

$$Precision = TP/TP + FP$$

7.7.7   RECALL (AFFECTABILITY)

It is the proportion of effectively anticipated positive perceptions to the all perceptions in real class-yes.

$$Recall = TP/TP+FN$$

7.7.8   F1 SCORE

It is the weighted normal of Precision and Recall, this score considers both FP and FN, it cannot be done as like accuracy, yet F1 is typically have more advantages than accuracy, particularly on the off chance that you have an uneven class distribution. Accuracy works well if FP and FN have comparable expenses. In the event that the expense of FP and FN are altogether different.

$$F1\ Score = 2*(Recall * Precision)/(Recall + Precision)$$

Computing precision and recall are very simple. Envision there are 100 positive cases among 10,000 cases. You need to foresee which ones are positive, and you pick 200 to have a superior possibility of getting a considerable lot of the 100 positive cases. You record the IDs of your expectations, and when you get the genuine outcomes, you total up how often you were correct or off-base. There are four different ways of being correct.

- TN: The scenario was –VE and anticipated –VE.
- TP: The scenario was +VE and anticipated +VE.
- FN: The scenario was +VE however anticipated –VE.
- FP: The scenario was-VE however anticipated +VE.

7.8   APPLICATIONS OF TEXT MINING (TM)

7.8.1   RESEARCH PAPERS CATEGORIZATION

In recent years, the volume of logical writing has become quickly bringing up an approaching issue about its stockpiling and association. Many research papers are often accessible just through the websites of the important logical diaries. It is a fundamental issue when distinctive classification codes are utilized so as to sort out these papers or when explicit categorization in a specific logical field is absent. This prompts superfluous confusions in the researchers' points that need to rapidly and effectively discover writing on a particular topic among the substantial measure of logical distributions.

At the same time, the research intrigue identified with the systems of NLP is developing since a significant part of the information they work with is unstructured and as plain text. So as to improve and robotize the process of sorting out and classifying logical papers we propose a methodology dependent on the innovation for NLP. This applies the strategies for managed machine learning and two explicit algorithms for text categorization-support vector machines (SVM) and Naive Bayes (NB). The proposed methodology classifies the logical writing as per its substance [9].

Logical research is a perplexing process, which can be related to a few key stages: research, composing, distributing, collaboration, assessment, and introduction. The most vital is the "research" organize, amid which the researchers get to know the research done as such far, look at the applicable logical papers and recognize potential research issues [53]. For them, it is likewise critical to discover the creators who have worked the most in a particular research field. Bibliographic and science-based databases are made so as to support the previously mentioned "research" organizes [24]. In light of the necessities of established researchers, the College of Financial aspects-Varna is likewise effectively chipping away at improving the quality and expanding the distributing action of the scholastic staff substance [9, 38].

The general methodology for the classification of logical writing:

1. Determining a classification conspire;
2. Perusing text documents;
3. Changing over into a structured configuration (Text preprocessing);
4. Classification.

7.8.2 HEALTHCARE

A few research thinks about have concentrated on the processing of textual information accessible in social insurance datasets. A concise outline of concentrates that feature the criticalness of textual data and its appropriateness in research settings is introduced here [24].

The motivation behind this investigation was to extract data identified with unfavorable occasions associated with focal venous catheter arrangement. Unfavorable occasions can be things, for example, contaminations, complexities from scattering, and pneumothorax (a fallen lung). Tests were directed utilizing every technique separately and after that utilizing them together on an example of records that had been physically looked into previously. The preliminaries utilizing the individual strategies were ineffective. The expression coordinating algorithm was not sufficiently explicit and the NLP framework was not sufficiently delicate. Researchers delivered positive forecast estimations of 6.4 and 6.2% individually. Be that as it may, when utilized together the outcomes were promising. They yielded a 72.0% affectability and 80.1% explicitness which are adequate qualities. This examination indicates the potential for utilizing NLP frameworks to robotize research data extraction [41].

Contrasted and strategies that are utilized by clinicians this framework fundamentally improved the positive prescient esteem. Concentrates, for example, these are particularly vital since a definitive objective is to move to a framework that can anticipate such events in future [24]. As of late, TM instruments have been used in medicinal services research, subsequently, TM of earlier master treatment can furnish doctors accessible if the need arises with an enhanced treatment plan. It can likewise prompt improvement of conventions to mitigate dissimilarity in treatment.

7.8.3   CUSTOMER ENLIGHTENMENT SYSTEMS

The customer relationship management (CRM) has seen new process called customer enlightening systems in the recent years as corporate feels that the basic information is available to its entire customer base. To enable the CRM to the next level of enlightening the customers, companies require exact information on the customer reviews on a particular product before addressing the issues [50]. Hence, it requires collecting all the related content from various sources which is leading to the importance of TM in this domain [38].

CA (content analysis) is the process of constructing a structure and selection of the units with analysis dependent on the objective of the examination. It utilizes the standards of "quantifiable" and "quantifiable" to plan classifications that can partition the broke down units' data content into an arrangement, chooses delegate data tests, and uses the classifications to evaluate and dissect the examples. Observational investigations in customary correspondence fields often utilize CA on a different style of ad [25].

Among customer relationship the executives consider, DTs are as often as possible utilized when examining customer portfolio the board. A fundamental purpose of DTs is the development of a spreading structure by classifying identified patterns [23].

7.8.4 FINANCIAL INSTITUTIONS

Recently, banks' dangers and particularly the administration thereof have formed into a noteworthy segment of the budgetary business. These incorporate credit, market, and liquidity chance, in particular, yet in addition operational, lawful, and numerous different dangers. It is in the money related organizations' enthusiasm to sufficiently represent its dangers, to stay dissolvable and to be profitable in the meantime [26].

Each bank imparts its dangers, the degree of these dangers, and the dangers' administration in the bank's hazard report. This report is an imperative instrument to decrease the information asymmetry between the bank and the perusers, for example, partners, bosses, or financial specialists.

It is likewise a device for outlining the previous year's improvements and to advice on the aftermath of emergencies. Adjacent to the open hazard revelation report, chiefs require the banks to report their dangers in an administrative divulgence report. Our research focuses on composed text, on subjective data, as opposed to quantitative numerical data. While administrative divulgence reports are institutionalized and don't convey much subjective information, hazard revelation reports are appropriate for subjective text analysis [26].

The report's casual structure gives banks a specific level of slack to report their dangers and hazard the board. Adjacent to these administrative necessities, banks have extra potential outcomes when investigating the association of their hazard to the executives. For instance, a few banks distribute "Hazard Maps" or "The board Cockpits." The depiction of hazard estimations, the collaboration of various customers' ventures, and economic figures, in particular, are definitely imparted in composed structure. Thusly, a bank's open hazard exposure report includes key information that isn't accessible as numerical data.

There are two fundamental approaches to process this information: either investigator intently perused each single report or we use machine processing to measure the text utilizing TM algorithms to extract the applicable information. Close perusing of hazard reports is costly in terms of time and workforce and is additionally inclined to mistake and subjectivity. In differentiate, we locate that robotized textual analysis performs rapidly and is objective to a substantial degree. Additionally, utilizing a board relapse, we acquaint a model with reliably evaluate how much a report satisfies the necessities, its quality. To the best of our knowledge, there has been no mechanized quality assessment of hazard reports to date [26].

7.8.5 SENTIMENTAL ANALYSIS

Sentiments are critical to most of the activities by human since they influence in our behavior it is very human nature to get other's opinion on decisions by self. In most of the situations, people try to implement expectation confirmation (EC) theorem during their decision-making process. Once the decision is made, it is normal for people to check the decision is a positive or negative type. For which we seek for others opinion in the form of conversation, questionnaire, survey, etc. [30].

The nature of the entertainment industry is taking a large revamp from traditional methods of Movie Theater to Home Theater with sophisticated on-demand services like DTH and other web services. All the programs are aired based on the viewership count that technically called TRP (target rating point). Channel operators are basically do opinion mining on the viewer comments for the particular program. This process has made an impact in the way the channels promote and air the programs. In today's scenario, opinions play a major role of a movie or program to become successful [32].

Sentimental analysis, otherwise called as opinion mining, fundamentally, is the process of evaluating the passionate incentive in a progression of words or text, to pick up a comprehension of the frames of mind, opinions, and feelings communicated [33]. Sentimental analysis can be connected to different areas, for example, eCommerce, banking, mining social media websites like Facebook, Twitter, etc. Utilizing assumption analysis and TM, associations can pick up purchaser knowledge from the reaction about their items and administrations [28]. This can be additionally used to consider customers' fulfillment with the administrations and if there should arise an occurrence of protests and issues, finding the conceivable purposes behind that. One of the uses of assumption analysis is proposal frameworks, for example, YouTube prescribes based on customer's likes, aversions, and remarks given by the client. In this paper, we broadly examine different TM and Sentimental analysis procedures connected to various regions in multilingual configuration and from various assets.

A notion analysis and TM system regularly incorporate the following subtasks: getting text data, data cleaning and preprocessing, data standardization, the transformation of text to machine intelligible vectors, features selection, lastly applying NLP and machine LAs [28]. Here it is presented a writing audit on ongoing patterns in TM and slant analysis. For example, shopper audit mining and application to the travel industry are the current effective applications. Topic displaying is effectively joined with estimation priors to produce topics and notion classes all the while. Emoticon and emoji assessments are incorporated into a large number of investigations to improve the accuracy of results, etc. [30].

If an online shopping portal wants to increase the sale for their business then it is not possible to implement without the knowledge of sentimental analysis, as there are 'N' number of products are available with the similar attributes. It is hard to stay in the business without understanding the emotions of the customer with the previous purchases on the portal. Most of the organization go through the reviews given by the customer about the product. If it is positive, then the work will be easy but if the customer given the reviews which are not likely as they expected then the organization should work as per the needs and wishes from the customers. It is the first and foremost step that every organization should be performing after the launch of the product. As per the flow of the diagram, initially, the views will be discussed, then the sentences will be extracted from the available reviews, followed by feature extraction, then the sentences will be analyzed whether the emotions are positive or negative, followed by sentence opinion, then the summarization of overall review, calculation of the review score which gives us the results (Figure 7.14).

This can be additionally used to consider customers' fulfillment with the administrations and used to take appropriate actions to increase the revenue by sentimental analysis.

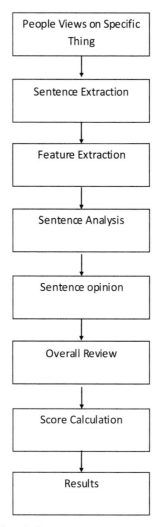

FIGURE 7.14   Sentimental analysis process.

7.9   CONCLUSION AND FUTURE WORK

As TM is the root of all the new technologies such as cloud computing, big data, data science, there are still lot of research work has been carried out by the researchers for the innovation of new methodology to reduce the work of humans. This chapter discusses about the available methodologies to perform and the most familiar algorithms, the applications are not limited to the classification of research papers, healthcare, CRM, education, it applies in all places where text has been stored as data. As it is already told that information is wealth, TM supports in using the information to the next level of decision making for business, result in analysis for the education sector and so on. The detailed TM algorithms are discussed such as decision trees, neural networks, discovering algorithms, differential evolution and so on. Applications such as healthcare, banks, social media, customer relationship and all are connected to the TM, these applications limited whereas the usage of TM are not limited to the above means. In simple terms, wherever the text has been stored as data in the database, it can be used extensively in taking decisions, predicting results, diagnosing patients, increasing sales and so on.

KEYWORDS

- **association rule mining**
- **customer relationship management**
- **data mining**
- **expectation confirmation**
- **hierarchical agglomerative clustering**
- **natural language processing**

REFERENCES

1. Allahyari, M., et al., (2017). "*A Brief Survey of Text Mining: Classification, Clustering and extraction techniques.*" arXiv preprint arXiv:1707.02919.
2. Jasmine, G. S. D., & Sundar, D., (2017). "A study of various text mining techniques." *International Journal of Advanced Networking and Applications (IJANA), 08*(05), 82–85.
3. Gupta, V., & Gurpreet, S. L., (2009). "A survey of text mining techniques and applications." *Journal of Emerging Technologies in Web Intelligence, 1*(1), 60–76.
4. Meena, P. B., & Radha, P. (2016). "A survey paper on text mining-techniques, applications, and issues." *IOSR Journal of Computer Engineering (IOSR-JCE)* (pp. 46–51). e-ISSN: 2278-0661, p-ISSN: 2278-8727.

5. Ghosh, S., Sudipta, R., & Bandyopadhyay, S., (2012). "A tutorial review on text mining algorithms." *International Journal of Advanced Research in Computer and Communication Engineering, 1*(4), 7.

6. Kannan, S., & Vairaprakash, G., (2014). "Preprocessing techniques for text mining." *International Journal of Computer Science and Communication Networks, 5*(1), 7–16.

7. Song, C., (2016). "Research of association rule algorithm based on data mining." *2016 IEEE International Conference on Big Data Analysis (ICBDA)*. IEEE.

8. Kumar, L., & Parul, K. B., (2013). "Text mining: Concepts, process and applications." *Journal of Global Research in Computer Science, 4*(3), 36–39.

9. Salloum, S. A., et al., (2018). "Using text mining techniques for extracting information from research articles." *Intelligent Natural Language Processing: Trends and Applications* (pp. 373–397). Springer, Cham.

10. Fang, C., Kesong, H., & Guilin, C., (2008). "An approach to sentence selection based text summarization." *Proceedings of IEEE TENCON02*, pp. 489–493.

11. Setu, M. N., Haiying, T., Jianhui, L., & Krishna, R. P., (2005). "Experiments on supervised learning algorithms for text categorization." *International Conference, IEEE Computer Society*, pp. 1–8.

12. Guihua, W., Gan, C., & Lijun, J., (2006). "Performing text categorization on manifold." *2006 IEEE International Conference on Systems, Man, and Cybernetics* (pp. 3872–3877). Taipei, Taiwan, IEEE.

13. Jian-Suo, X., (2007). "TCBPLK: A new method of text categorization." *Proceedings of the Sixth International Conference on Machine Learning and Cybernetics* (pp. 3889–3892). Hong Kong, IEEE.

14. Ming, Z., Jianli, W., & Guanjun, F., (2008). "Research on application of improved text cluster algorithm in intelligent QA system." In: *Proceedings of the Second International Conference on Genetic and Evolutionary Computing* (pp. 463–466). China, IEEE Computer Society.

15. Xi Quan, Y., Di Na, G., Xue, Y. C., & Jian, Y. Z., (2008). "Research on ontology-based text clustering." In: *Third International Workshop on Semantic Media Adaptation and Personalization* (pp. 141–146). China, IEEE Computer Society.

16. Zhou, N., Wu, J., Wang, B., & Zhang, S., (2008). "A visualization model for information resources management." In: *12th International Conference Information Visualization* (pp. 57–62). China, IEEE.

17. Jignashu, P., & Narasimha, M. M., (2002). "Adapting question answering techniques to the web." *Proceedings of the Language Engineering Conference*. India, IEEE Computer Society.

18. Emilio, S., Davide, B., Sergio, G., Lluis, H., & David, G., (2006). "Spoken QA based on a passage retrieval engine." In: *Proceedings of IEEE International Conference* (pp. 62–65). Spain.

19. Li, G., Elizabeth, C., & Song, H., (2005). "Powerful tool to expand business intelligence: Text mining." *Proceedings of World Academy of Science, Engineering, and Technology* (Vol. 8, pp. 110–115).

20. Rajender, S. C., (2008). "Extraction transformation loading: A road to data warehouse." In: *2nd National Conference Mathematical Techniques: Emerging Paradigms for Electronics and IT Industries* (pp. 384–388). India.

21. Dion, H. G., & Rebecca, P. A., (2007). "An introduction to association rule mining: An application in counseling and help seeking behavior of adolescents." *Journal of Behavior Research Methods, 39*(2), 259–266, Singapore.
22. Kanya, N., & Geetha, S., (2007). "Information extraction: A text mining approach." *IET-UK International Conference on Information and Communication Technology in Electrical Sciences* (pp. 1111–1118). IEEE, Dr. M.G.R. University, Chennai, Tamil Nadu, India.
23. Shantanu, G., & Shourya, R., (2008). *"Text to Intelligence: Building and Deploying a Text Mining Solution in the Services Industry for Customer Satisfaction Analysis"* (pp. 441–448). IEEE.
24. Raja, U., et al., (2008). "Text mining in healthcare. Applications and opportunities." *J. Healthc. Inf. Manag., 22*(3), 52–56.
25. Chang, C. W., Chin-Tsai, L., & Lian-Qing, W., (2009). "Mining the text information to optimizing the customer relationship management." *Expert Systems with Applications, 36*(2), 1433–1443.
26. Fritz, D., & Eugen, T., (2018). "Text mining and reporting quality in German banks-A co-occurrence and Sentiment Analysis." *Universal Journal of Accounting and Finance, 6*(2), 54–81.
27. Jotheeswaran, J., & Koteeswaran, S., (2015). "Decision tree based feature selection and multilayer perceptron for sentiment analysis." *Journal of Engineering and Applied Sciences, 10*(14), 5883–5894.
28. Swathi, R., Sangeet, S., Barkha, B., & Gaurav, G., (2018). "Sentiment analysis using text mining: A review." *International Journal on Data Science and Technology, 4*(2), 49–53. http://www.sciencepublishinggroup.com/j/ijdst (accessed on 26 February 2020). doi: 10.11648/j.ijdst.20180402.12.
29. Jotheeswaran, J., & Kumaraswamy, Y. S., (2013). "Opinion mining using decision tree based feature selection through Manhattan hierarchical cluster measure." *Journal of Theoretical and Applied Information Technology, 58*(1).
30. Jotheeswaran, J., & Koteeswaran, S., (2016). "Feature selection using random forest method for sentiment analysis." *Indian Journal of Science and Technology, 9*(3), 1–7.
31. Mostafa, M. M., (2013). "More than words: Social networks' text mining for consumer brand sentiments." *Expert Systems with Applications, 40*(10), 4241–4251.
32. Jeevanandam, J., D. R., & Kumaraswamy, Y. S., (2013). "Opinion mining using decision tree based feature selection through Manhattan hierarchical cluster measure." *Journal of Theoretical and Applied Information Technology, 58*(1).
33. Shaikh, T., & Deepa, D. (2016). *"Feature Selection Methods in Sentiment Analysis and Sentiment Classification of Amazon Product Reviews."*
34. Jotheeswaran, J., & Koteeswaran, S., (2016). "Mining medical opinions using hybrid genetic algorithm—neural network." *Journal of Medical Imaging and Health Informatics, 6*(8), 1925–1928.
35. Jeevanandam, J., & Koteeswaran, S. (2015). *"Sentiment Analysis: A Survey of Current Research and Techniques."*
36. Wong, P. C., Paul, W., & Jim, T., (1999). "Visualizing association rules for text mining." *Proceedings 1999 IEEE Symposium on Information Visualization (InfoVis' 99).* IEEE.
37. Nasukawa, T., & Tohru, N., (2001). "Text analysis and knowledge mining system." *IBM Systems Journal, 40*(4), 967–984.

38. Berry, M. W., & Jacob, K., (2010). *Text Mining: Applications and Theory*. John Wiley & Sons.
39. Jotheeswaran, J., Loganathan, R., & Madhu, S. B., (2012). "Feature reduction using principal component analysis for opinion mining." *International Journal of Computer Science and Telecommunications, 3*(5), 118–121.
40. Billheimer, D. D., et al., (2003). *"Method and System for Text Mining Using Multidimensional Subspaces."* U.S. Patent No. 6,611,825.
41. Zweigenbaum, P., et al., (2007). "Frontiers of biomedical text mining: Current progress." *Briefings in Bioinformatics, 8*(5), 358–375.
42. Kao, A., & Steve, R. P., (2007). *Natural Language Processing and text Mining.* Springer Science & Business Media.
43. Spasic, I., et al., (2005). "Text mining and ontology's in biomedicine: Making sense of raw text." *Briefings in Bioinformatics, 6*(3), 239–251.
44. Nasukawa, T., & Tohru, N., (2001). "Text analysis and knowledge mining system." *IBM Systems Journal, 40*(4), 967–984.
45. Srivastava, A. N., & Mehran, S., (2009). *Text Mining: Classification, Clustering, and Applications*. Chapman and Hall/CRC.
46. Nahm, U. Y., & Raymond, J. M., (2002). "Text mining with information extraction." *Proceedings of the AAAI 2002 Spring Symposium on Mining Answers from Texts and Knowledge Bases*. Stanford CA.
47. Ghani, R., et al., (2006). "Text mining for product attributes extraction." *ACM SIGKDD Explorations Newsletter, 8*(1), 41–48.
48. Krallinger, M., Alfonso, V., & Lynette, H., (2008). "Linking genes to literature: Text mining, information extraction, and retrieval applications for biology." *Genome Biology, 9*(2), S8.
49. Adeva, J. J. G., & Juan, M. P. A., (2007). "Intrusion detection in web applications using text mining." *Engineering Applications of Artificial Intelligence, 20*(4), 555–566.
50. Nassirtoussi, A. K., et al., (2014). "Text mining for market prediction: A systematic review." *Expert Systems with Applications, 41*(16), 7653–7670.
51. Kosala, R., & Hendrik, B., (2000). "Web mining research: A survey." *ACM Sigkdd Explorations Newsletter, 2*(1), 1–15.
52. Aggarwal, C. C., & Cheng, X. Z., (2012). "An introduction to text mining." *Mining Text Data* (pp. 1–10). Springer, Boston, MA.
53. Abbe, A., et al., (2016). "Text mining applications in psychiatry: A systematic literature review." *International Journal of Methods in Psychiatric Research, 25*(2), 86–100.
54. Raja, U., et al., (2008). "Text mining in healthcare. Applications and opportunities." *J. Healthc. Inf. Manag., 22*(3), 52–56.
55. Jotheeswaran, J., Loganathan, R. and Madhu Sudhanan, B., 2012. Feature reduction using principal component analysis for opinion mining. International Journal of Computer Science and Telecommunications, 3(5), pp.118-121.
56. Mostafa, M.M., 2013. More than words: Social networks' text mining for consumer brand sentiments. Expert Systems with Applications, 40(10), pp.4241-4251.
57. Adeva, J.J.G. and Atxa, J.M.P., 2007. Intrusion detection in web applications using text mining. Engineering Applications of Artificial Intelligence, 20(4), pp.555-566.

CHAPTER 8

A Brief Overview of Natural Language Processing and Artificial Intelligence

SUSHREE BIBHUPRADA B. PRIYADARSHINI,[1]
AMIYA BHUSAN BAGJADAB,[2] and BROJO KISHORE MISHRA[3]

[1]*Institute of Technical Education and Research, Siksha 'O' Anusandhan (Deemed to be University), Bhubaneswar, India,*
E-mail: bimalabibhuprada@gmail.com

[2]*Sambalpur University of Information Technology, Burla, India,*
E-mail: amiya7bhusan7@gmail.com

[3]*Gandhi Institute of Engineering and Technology (GIET), Gunupur, India, E-mail: brojokishoremishra@gmail.com*

ABSTRACT

In the modern era of technological peregrination, artificial intelligence (AI) is one of the most interesting fields of research and language is the most compelling manifestation of intelligence. In this context, natural language processing (NLP) stands as a popular field of research in current days. Basically, NLP is employed to allow the machines of how humans being speaks and basically aimed at analyzing the text. Moreover, NLP is normally employed for text mining (TM), machine translation, automated question answering, etc. Such human-computer interaction enables real-world applications such as sentiment analysis, automatic text summarization, topic extraction, stemming, machine translation, etc. This chapter further discusses the two basic aspects like natural language understanding (NLU) and natural language generation (NLG) that deal with NLU and generation respectively. Further, the current chapter throws light on various aspects of text processing like: morphological analysis, syntax analysis, semantic analysis, lexical analysis, etc.

8.1 INTRODUCTION TO ARTIFICIAL INTELLIGENCE (AI) AND NATURAL LANGUAGE PROCESSING (NLP)

Artificial intelligence (AI) is a field of computer science that focuses on the generation of intelligent machines that behave and react like humans. Various activities are performed with computers employing AI including Speech recognition, natural language processing (NLP), etc. NLP represents a sub-field of AI which deals with the interaction between the human language and computer. Specifically, it deals with how to program the computer to analyze and process a huge amount of data [1].

The history of NLP initiated in 1950. However, a notably successful NLP system got developed in the year 1960. Up to 1980, most NLP systems were based on sets of handwritten rules. Further, NLP can be classified into two broad categories such as: natural language understanding (NLU) and natural language generation (NLG) as portrayed in Figure 8.1. The term NLU means the identification of the desired semantic form various possible semantics those can be derived from a natural language expression. NLG involves the process of generating the natural language [1].

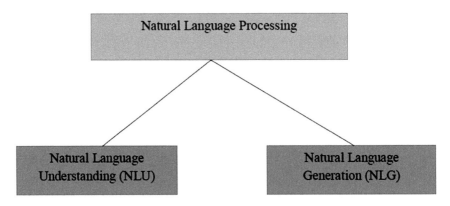

FIGURE 8.1   Types of natural language processing (NLP).

8.1.1   TYPES OF LANGUAGE ANALYSIS

The language analysis is classified into the types as indicated in Figure 8.2.

8.1.1.1 *MORPHOLOGICAL ANALYSIS*

This is the process of finding the root word (i.e. Base word) along with its gender, number, person, etc. In this context, gender can be classified as male and female. The number can be singular or plural [2]. Further, the person can be first, second or third person. Consider Figure 8.3. Consider the plural phrase "cats," where the root word is the cat, which represents a third person. Similarly, consider the statement "He is going." as shown in Figure 8.4. Here the root word is "go" and "going" represents the verb. Similarly, consider the statement: "I am going to school": "I" represents the subject which is a noun. The preposition is the "to." The school represents the noun.

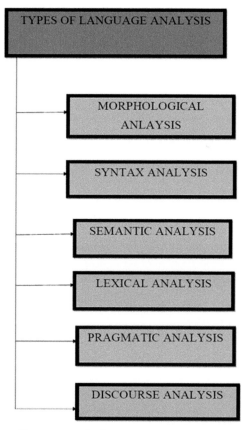

FIGURE 8.2 Types of language analysis.

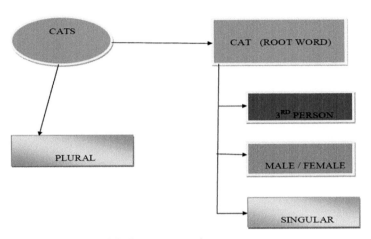

FIGURE 8.3   A scenario of finding a root word.

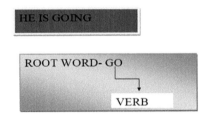

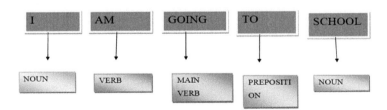

FIGURE 8.4   A scenario of morphological analysis.

8.1.1.2   PRAGMATIC ANALYSIS

It represents the process of extracting information from text. Particularly, it represents the portion that emphasizes on taking a structured set of text what figuring out the actual meaning [3].

8.1.1.3 DISCOURSE ANALYSIS

Discourse analysis can be depicted as the analysis of language "beyond the sentence" [4]. which contradicts with categories of analysis more specially recent linguistics, that are primarily associated to the study of grammar along with those of smaller bits of language, incorporating sounds (phonetics and phonology), parts of words (morphology), meaning (semantics), as well as the sequence of words in sentences (syntax). Discourse analysts study larger chunks of language as they flow at a time.

Various discourse analysts consider the larger discourse matter so as to understand how it hampers the meaning of concerned sentence. For instance, Charles Fillmore points out that two sentences considered duo at a time as a single one can possess us to consider two independent signs at a river: "Please use the washroom, not the river," says one person. The other tells, "Pool for merely the members "If one regards each one independently, they appear to be solely reasonable. However, considering both at a time as a single discourse makes one go back and revise the interpretation of the first one after considering the second one.

8.1.1.4 SYNTAX ANALYSIS

A syntax analyzer i.e., parser gathers the input through a lexical analyzer in the form of "token streams." In this context, the parser analyzes the respective source code (token stream) against the production rules for checking any errors in the code. The output of this phase represents a parse tree. The syntax analysis checks whether a statement is syntactically right or not [5].

8.1.1.5 SEMANTIC ANALYSIS

Semantic analysis means the task of ensuring that some statements and declaration of programs are semantically right. That means the meaning is clear and consistent with the way in which the data types, as well as the control structures, are supposed to be employed [6].

Basically, the semantics of a language affords the meaning to the constructs such as tokens; interpret symbols, their types, and the concerned relationship. Various types of semantic errors include: type mismatch, undeclared variable, multiple declaration of a variable in a scope, actual and formal parameter mismatch, and reserved identifier misuse.

8.1.1.6   LEXICAL ANALYSIS

Lexical analysis represents the initial phase of the compiler, which collects the modified source code from language preprocessors which are written in the form of sentences. The lexical analyzer divides such syntaxes into a series of tokens by eliminating any white spaces or comments residing in the concerned source code. This process is also known as lexing or tokenization. In this context, a program that performs lexical analysis is the lexer or tokenizer [7].

8.2   WORD-SENSE DISAMBIGUATION (WSD)

In computational linguistics, Word Sense Ambiguation (WSD) represents an open problem that identifies the sense of a word that is being used in any sentence. The human brain is completely efficient at word-sense disambiguation (WSD). It has been a significant ultimatum in developing the capability of any computer to carry out NLP and machine learning [8]. Various dictionary-based methods/strategies are employed along with supervised machine learning to tackle such a task.

For example, let us consider Figure 8.5 where the term "Bank" is having double meaning like "riverbank" or the "financial bank." Likewise, the word "book" can be employed both as a noun as well as a verb as illustrated in Figure 8.5.

In this context, there exist four conventional approaches pertaining to WSD:

8.2.1   DICTIONARY AND KNOWLEDGE-BASED METHOD

Such methods basically focus on the dictionary and lexical information bases in lieu of employing any corpus proof. The Lesk algorithms represent [10] the seminal dictionary-based strategy, which is based on the hypothesis that words are employed together in any text are concerned with one another as well as the relation will be get tracked in the definitions of the words and their concerned senses [8].

Further, two (or more) words are disambiguated by choosing the pair of dictionary senses with the greatest word overlap in the corresponding dictionary definitions. For example, while disambiguating the words in "pine cone," the definitions of the appropriate senses both incorporate the words evergreen and tree.

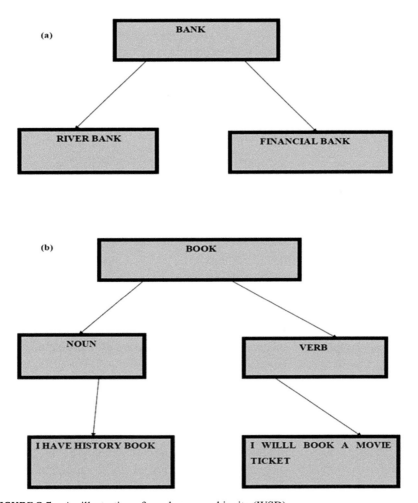

FIGURE 8.5   An illustration of word sense ambiguity (WSD).

8.2.2 SUPERVISED METHODS

Supervised strategies are based on the presumption that the context can afford ample proof on its own to disambiguate words. In this context, perhaps, every machine learning algorithm (LA) has been applied to WSD, incorporating related strategies such as [8] feature selection, parameter optimization, and ensemble learning.

Further, support vector machines (SVM), as well as memory-based learning, have been shown to be the most successful strategies to date, probably since they can cope with the high-dimensionality of the feature space. Further, such supervised methods [8] get subjected to a new knowledge acquisition bottleneck as they rely on substantial amounts of manually sense-tagged corpora for training, that are laborious and expensive to form.

8.2.3   SEMI-SUPERVISED METHODS

Due to the lack of training data, various WSD strategies employ semi-supervised learning that permits both labeled as well as unlabeled data. The arowsky strategy was an early illustration of such a strategy that employs the 'One sense per collocation' and the 'One sense per discourse' properties of human languages for WSD. From speculation, words tend to show merely one sense in most given discourse and in a particular collocation.

Further, the bootstrapping methodology initiates from a little amount of seed data for every word: either manually tagged training illustrations or a small number of decision rules. Moreover, the seeds are engaged to train an initial classifier, employing any supervised strategy. This classifier is then used on the untagged portion of the corpus to extract a larger training set, in which merely the most confident classifications get considered.

Such process repeats, every novel classifier being trained on a successively larger training corpus, till the whole corpus gets consumed, or till a given maximum number of iterations are reached. Such strategies have the potential to assist in the adaptation of supervised models to various domains [8].

8.2.4   UNSUPERVISED METHODS

Unsupervised learning represents the greatest ultimatum for WSD investigators. The assumption is that similar senses prevail in similar types of contexts, and therefore, senses can get induced from the text by considering word occurrences using the similarity in the context. Further, new occurrences of the word can get classified into the closest induced clusters/senses. Further, word sense induction strategies can be tested and compared within any application prospect. Further, word sense induction improves [8] Web search outcome clustering by enhancing the quality of resultant clusters conjointly with the degree diversification of result lists.

8.2.5   OTHER APPROACHES

Other strategies differ in their methods used [8]:

- Disambiguation based on operational semantics of default logic.
- Domain-driven disambiguation.
- Identification of dominant word senses.
- WSD using cross-lingual evidence.
- WSD solution in John Ball's language-independent NLU combining Patom

8.3   STEMMING

Stemming refers to the process of diminishing a word to its word stem which gets affixed to the suffixes and prefixes or to the roots of the words called the lemma. Stemming is very crucial in NLP as well as NLU (Figure 8.6).

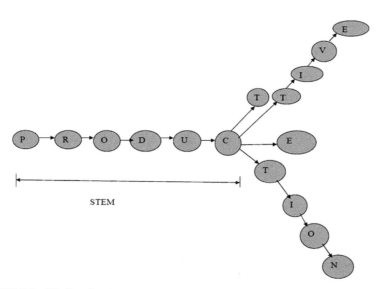

FIGURE 8.6   Finding the stem.

Basically, it refers to a part of linguistics studies in morphology as well as AI for information extraction (IE) and retrieval. The process of stemming conjointly with AI extracts meaningful information from vast sources such as the internet as well as big data. In this context, additional forms of words related to a subject need to be investigated to get the best results. Stemming represents a part of internet search engines. Various examples of stemming algorithms incorporate [9]:

1. Lookups in tables of inflected forms of words that need all inflected forms are listed.
2. **Suffix:** The algorithms recognize known suffixes on inflected words and discard them.
3. **Lemmatization:** This algorithm gathers all inflected forms of a word so as to break them down to their root dictionary form or lemma. Words get broken down into a part of speech considering the way of the rules of grammar.
4. **Stochastic Models:** Such an algorithm earns from tables of inflected forms of words. Through understanding suffixes, as well as the rules through which they are applied, an algorithm can stem novel words.

Let us consider the words like production, producer, as well as product. All these three words are framed from the stem "produce" as shown in Figure 8.6.

8.4 VARIOUS PILLARS OF NLP

There exist four basic foundations pertaining to neuro-linguistic programming [10]. They are as given in the subsections.

8.4.1 SENSORY AWARENESS

At times, when we consider someone else's home, we mark that the colors, smells, textures, as well as sounds, are completely distinct from our own. Similarly, we can observe various types of distinct clothes while we go for shopping in a big mall. In this connection, neuro-linguistic programming enables us to notice that our world is much richer while focusing on own senses completely [10].

8.4.2   RAPPORT

NLP affords a crucial gift in framing relationships with different people. Rapport can be depicted as connecting with others quicker. Generating rapport frames of trust from other people. Such rapport can be built easily through understanding modality preferences, eye accessing cues, and predicates [10].

8.4.3   OUTCOME-BASED THINKING PROCESS

An outcome refers to the goal for accomplishing anything. Outcome is concerned with thinking regarding what we desire, in contrast to getting stuck in any kind of negative thinking. The fundamentals of outcome strategy may assist in making the best decision out of a number of choices [10].

8.4.4   FLEXIBILITY IN BEHAVIOR

Behavioral flexibility refers to being able to carry out something in a different way if the current way is not working. Flexibility in behavior is a key aspect in practicing of NLP. The NLP assists in finding fresh perspectives and to frame such habits into action [10].

8.5   NLP TERMINOLOGY

The crucial terminologies used in NLP are as follows in the subsections [2, 10, 11].

8.5.1   MORPHOLOGY

The term Morphology refers to the study of the formation of words from primitive meaningful components. Further, morphology also considers the parts of speech, stress, intonation, and the methods through which a word's pronunciation, as well as meaning, varies. Further, it should be noted that morphology differs from that of morphological typology, which represents the classification of languages based on the use of words. Similarly, morphology is different from that of lexicology, which represents the study of words and the way they can make up the language vocabulary [2, 10–12].

8.5.2   MORPHEME

Morpheme indicates the meaning in a language. In other words, morpheme refers to the morphological unit of a language that cannot be further segregated. In other words, it represents the smallest grammatical unit in a language. It is different from the word. The underlying difference is that morpheme may or may not stand alone whereas word can freely stand. The concerned linguistic field of study is morphology.

8.5.3   SYNTAX

Syntax refers to the arrangement of words to frame a sentence. Moreover, it is further concerned with the structural features of corresponding words in the respective sentence.

8.5.4   SEMANTICS

Semantics refers to the meaning of words and the way of combining words into meaningful phrases as well as sentences.

8.5.5   PHONOLOGY

Phonology refers to the study of organizing sound systematically. In other words, it refers to the branch of linguistics that deals with systems of sounds within a language or between several languages.

8.5.6   PRAGMATICS

Pragmatics refers to usage and understanding of sentences in distinct situations and the way the interpretation of the sentence gets affected. It encompasses speech act theory, conversational implicature, sociology, linguistics, as well as anthropology, etc.

8.5.7 DISCOURSE

Discourse deals with how the immediately preceding sentence can affect the interpretation of the coming sentence.

8.5.8 WORLD KNOWLEDGE

It incorporates general knowledge regarding the world.

8.6 CONCLUSIONS

This chapter discusses the basic concepts of NLP which stand as a subfield to AI. The chapter throws light on the various types of language analysis and WSD and the associated methodologies. The process of stemming along with the pillars of NLP gets discussed. Further, for providing a brief overview, we have discussed the various crucial terminologies employed in the context of NLP in the domain of AI.

KEYWORDS

- **artificial intelligence**
- **natural language generation**
- **natural language processing**
- **natural language understanding**
- **pragmatics**
- **word-sense disambiguation**

REFERENCES

1. https://searchbusinessanalytics.techtarget.com/definition/natural-language-processing-NLP (accessed on 26 February 2020).
2. https://www.google.com/ (accessed on 26 February 2020).

3. https://www.quora.com/What-is-pragmatic-analysis-in-NLP (accessed on 26 February 2020).
4. https://www.linguisticsociety.org/resource/discourse-analysis-what-speakers-do-conversation (accessed on 26 February 2020).
5. https://www.tutorialspoint.com/compiler_design/compiler_design_syntax_analysis.htm (accessed on 26 February 2020).
6. https://www.tutorialspoint.com/compiler_design/compiler_design_semantic_analysis.htm (accessed on 26 February 2020).
7. https://en.wikipedia.org/wiki/Lexical_analysis (accessed on 26 February 2020).
8. https://en.wikipedia.org/wiki/Word-sense_disambiguation (accessed on 26 February 2020).
9. https://en.wikipedia.org/wiki/Stemming (accessed on 26 February 2020).
10. https://inlpcenter.org/what-is-neuro-linguistic-programming-nlp/ (accessed on 26 February 2020).
11. https://www.tutorialspoint.com/artificial_intelligence/artificial_intelligence_natural_language_processing.htm (accessed on 26 February 2020).
12. Braden, H., Paroma, V., Stephanie, W., Martin, B., Percy, L., & Christopher, R., (2018). *Training Classifiers with Natural Language Explanations.*

CHAPTER 9

Use of Machine Learning and a Natural Language Processing Approach for Detecting Phishing Attacks

CHANDRAKANTA MAHANTY, DEVPRIYA PANDA, and
BROJO KISHORE MISHRA

Department of CSE & IT, GIET University, Gunupur, Odisha, India

ABSTRACT

Phishing attacks are today a major threat to the security of the systems. And also, there is no foolproof protective system against these attacks. Phishing is one of the different cybercrimes. In this, the person interested to attack behaves like another existing individual/organization and he uses e-mails or similar types of techniques. It also can be described as appearing as a person or organization that one can trust and then to acquire certain private and significant data without the knowledge of the person concerned, such as sign-in credentials and bank-related card details, for fraudulent reasons. Even from chatting to banking, a huge community uses the online services through online transactions. Phishing attacks are happened by the execution of certain actions such as mouse clicking and hovering on malicious URLs. The attacker may also use phishing, by providing links which are malicious in nature through emails that can be used to capture login credentials, victim account information, etc. Therefore, we have to enhance the security mechanism.

In this book chapter paper, types of phishing attacks are explained. It also focuses on the anti-phishing URL tool which is used to prevent phishing attacks. The main objective of this book chapter is to explain initially the characteristics of phishing attacks. There are some uniqueness and patterns associated with the websites which are used for phishing.

There properties can be used to detect phishing. Then these attacks are detected by a hybrid machine learning model. The system has been implemented by examining the URLs used in phishing attacks with some extracted features before opening them. Some natural language processing (NLP) techniques are used in the proposed machine learning system. These techniques are used for analyzing the text semantically to detect malicious intentions which indicate phishing attacks. In order to identify the websites for their legitimacy, some machine learning algorithms (LAs) are also discussed in this book chapter. It also focuses on the Naive Bayes (NB) classifier, K-Means clustering to calculate the possibility of the website as valid phish or invalid phish.

9.1 INTRODUCTION

Phishing is defined by the intended recipients as fraudulently acquiring confidential data and misusing such data. Emails are generally used to carry out this attack. Emails which appear like sent from a legitimate source such as a bank, Credit Card Company or email provider is a phishing example. In general, updating accounts requires private data such as credit card numbers or passwords. These emails have a link to the URL which misleads users to a different website. Actually, it's a fake or altered website. When any person visits any website like this, the website misguides them and requests for information regarding credentials to access user accounts or bank accounts then transmits that to the person who is exercising the phishing attack [1, 2]. This is mostly used to gain knowledge of information about someone's credentials such as a password or credit card pin. In phishing, computer users are directed to fake sites by the emails prepared to disguise as official communications.

9.1.1 PHISHING ATTACKS

Researchers find phishing attacks as one of the various threats that can use email to establish itself [3]. Attackers are usually portrayed themselves as social websites, banks, IT department administrators, or popular sites for shopping. These emails may indeed attract us to click on hyperlinks to start downloads of malware, or provide private credentials on a website that looks like a legitimate one but actually a malicious one. Metadata of emails are generally used by most of the detection approaches for phishing. Metadata are email-related data that is unrelated to the semantic meaning of the text message. The URLs contained within the message are examined by several approaches. Different approaches are available for detecting phishing. Some of them search for precise terms in each sentence to evaluate the text [9, 10]. In phishing, a deceptive message which seems to be initiated from a legitimate source is sent. E-mail is usually used for the same. The goal is to thieve personal information like information regarding credit card and login credentials or to install malware on the user's system. Phishing is a frequent cyber-attack which everybody needs to learn for protection Phishing begins with a bogus email or other such transmission aimed at attracting a target. The communication seems to have originated from a trustworthy source in this sort of attack. If the victim is deceived by the attacker, then he/she is encouraged to deliver private data in a fake web site. Or malware are sometimes installed to the destination system. Attackers provide misleading financial gain by obtaining credit card information from their victim or other personal information. Phishing emails are sometimes used to get hold of login details or other employee information to be used for a highly structured complex attack against a particular organization.

Social networking or other unrestricted information resources such as Facebook may be used by attackers to collect contextual information on the victim's history, interests, and activities [36]. Using these prior information attackers can get relate to the information such as names, designations, and e-mail ids of potential victims. This collected information may be exploited to forge a credible e-mail. Fraudulent hyperlinks and attachments can be sent through emails to start this type of attacks. In this type of attacks, Facebook's most popular feeds were identified as the most vulnerable environment. The same is one the mostly used phishing environment that uses clickable links. While phishing assaults are made, they are frequently utilized for doubtful news, for example, those made around significant occasions and commemorations. In phishing, the typical case is that we receive a communication which we think is sent by a known individual or association. Then this cyber assault is completed by means of a malignant record infusion that incorporates phishing programming or through connections to vindictive sites. In either case, the objective is to guide the user to a pernicious site to introduce malevolent programming software on the gadget or to uncover individual and budgetary data, for example, passwords, account IDs, or card details. Figure 9.1 illustrates the phishing attacks process.

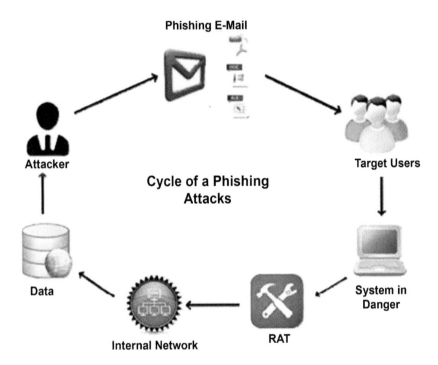

FIGURE 9.1   The processing cycle of phishing attacks.

9.1.2   *DIFFERENT PHISHING ATTACKS*

1. **Phishing Attacks Based on E-Mail:** The phishing attacker, in this case, eventually relies on the reaction of the person in question. How this attack is accomplished [4]? The appropriate response might be basic, yet the results can be distinctive in the different kind of methodology being utilized by the aggressor. Presently discussing the execution of such assaults, phishing assaults based on emails are conveyed by sending an email to the intended individual containing a hyperlink to a counterfeit site. Albeit this type attacks might be effective, the achievement status in this kind of assault from the assailant perspective is lower on the grounds that

numerous clients have learned not to send touchy data by means of messages.

2. **Phishing Attacks Based on Exploit:** Some of the phishing assaults are in fact follows all new methods and uses most common vulnerabilities in browser to introduce pernicious (virus or malware) software that accumulates delicate data of the unfortunate user [5]. For instance, an assailant can change the proxy settings to play the man-in-the-center attack so that all the traffic from the client will go through his server. For lessening phishing assaults based on exploit and other security dangers that contain viruses, worms, Trojans, and spyware, a producer of the browser should ensure that their product is without bug and the clients have introduced the most recent version of security features [5].

9.1.3 GENERAL IDEA REGARDING PHISHING DETECTION TOOLS

1. **Netcraft Toolbar:** To determine a website's legitimacy, the Netcraft Anti-Phishing Toolbar utilizes quite a few techniques. The Netcraft toolbar detects regularly websites with URLs containing characters that are meaningless. It can also be used to extract location of hosting, also the website country for the fraudulent URLs [5]. This tool maintains a black list of websites. If a user attempts to access any of them then it warns the user using a pop-up message.

2. **Anti-Phishing Toolbar:** Mozilla Firefox includes a module called Anti-Phishing. It helps in securing novice clients from getting affected by phishing assaults which use spoofing. It normally keeps track of the important information of the client and doesn't allow these information to be transmitted to a website that it considers as not a trust one [4]. The advancement of Anti-Phishing was roused via mechanized form-filler applications [5]. This application checks if the website has a safe association with SSL certificate. On the off chance that a website does not have secure association with SSL certificate, at that point the instrument creates a notice saying that don't enter delicate data on the grounds that the page don't have secure association with SSL certificate.

   A developer of a browser must prepare the product bug free and the clients who are using that product are to be introduced the most recent version of security patches. It should be done in order to reduce phishing assaults based on exploit and other security threats that contain viruses, worms, Trojans, and spyware [5].

3. **BitDefender Traffic Light Toolbar:** The BitDefender tool utilizes a blend of heuristics and blacklists. Green, red, and yellow are the three different modes used in this tool. If the user tries to visit a webpage that isn't unlawful or a phishing website then a symbol containing a green flag is shown. If the website is listed as a phishing website then red flag is shown. If it cannot categorize the website then it shows a yellow flag. If a recently known phishing site is encountered, it tries to block it and a pop up comes up which the client can use for superseding the block [6]. In Bitdefender Traffic Light toolbar, the client will dependably be educated about malware and deceitful sites inside search results. If a website comes with trackers then this tool can also identify it even the area of the trackers. There may be few pages which follow and break down client's browsing habits. Bitdefender Traffic Light demonstrates the client which pages uses this kind of code snippets and records them [7].

4. **SpoofGuard Toolbar:** SpoofGuard is a phishing toolbar created at Stanford University. It pursues different guidelines to recognize phishing web pages. The area of the present site is checked by the toolbar. The name is then compared with the sites regularly visited by the client. Then it investigates the complete URL to find out the port numbers which are not cleared or of nonstandard. Every website visited by the client is assessed by this tool and a score it decided for it. It also records these scores for every website [6]. Score for every site is figured as weighted total of the outcomes having a place with each arrangement of heuristics [5]. It likewise gives an alternative to change the loads for each arrangement of weight by the client. It shows a red symbol if the score determined outperforms a specific edge, cautioning clients that the site is a phishing site. On the off chance, that the score is adjacent or equivalent to the limit, the symbol turns yellow demonstrating that assurance of the site was ineffective. A site is accounted for safe with a green symbol in the toolbar if none of the heuristics were activated [6].

5. **PhishDetector Toolbar:** An extension to the Google Chrome is used to fake financial websites. This extension is known as PhishDetector.

It is a standard-based framework that examines the website page substance to recognize phishing assaults [8]. This toolbar distinguishes internet banking tricks all the more rapidly when contrasted with different tools and with zero false negative value. To shield a client from getting to the false financial site it is highly suggested to introduce this extension in the browser [8]. This tool distinguishes a legitimate banking webpage as sheltered by the substance accessible on the website page. It identifies a phishing website dependent on the content review on the site page.

It has been observed that when it comes to identify the attacks Anti-Phishing tool outperforms others by identifying around 94.32%. BitDefender Traffic Light follows it with 93.92%. Others perform in the following sequence-SoofGuard (82.16%), Netcraft (77.12%) and Phish-Detector (58.48%) [41].

9.1.4 PHISHING DETECTION ALGORITHM

A sentence is viewed as malicious in the event that it asks touchy data or directs an act of activity that may uncover individual data. Natural language processing (NLP) procedures are connected to parse each sentence and distinguish the semantic jobs of essential words in the sentence in connection to the predicate.

The algorithm for detecting phishing emails is shown below:

```
define SEAHound(text)
bad, urgent, generic = False
      for each sentence s in text
            bad | = BadQuestion(s) OR BadCommand(s)
            urgent | = UrgentTone(s)
            generic | = GenericGreeting(s)
            link = LinkAnalysis(s)
            if link
                  return True
      if majority(bad, urgent, generic)
            return True
      return False
```

The SEAHound algorithm [37] in the above assesses each sentence (steps 3–9) to decide whether it shows four qualities: 1) malicious question/command (line 4), 2) urgent tone (line 5), 3) generic greeting (line 6), and 4) malicious URL link (line 7). An email is viewed as malicious if a malignant connection is discovered (lines 8–9) or if somewhere around 2 of the staying three attributes are found in the email (lines 10–11). The LinkAnalysis step which checks the legitimacy of a URL is performed utilizing the Netcraft Anti-Phishing Toolbar. This Anti-Phishing toolbar is a viable and powerful toolbar [11]. Despite the fact that the connection examination given by Netcraft is successful, it is restricted to URL investigation, so it can't distinguish social engineering attacks which do exclude URL links.

Here the author uses the Scikit-learn Python library [12] and MultinomialNB() function. SEAHound delivers an expectation mark for every (verb-direct object) pair, and creates a certainty score (0 and 1) for the forecast. They utilized 1000 phishing and non-phishing email set from Nazario [13] and Enron Corpus [14] as preparing set respectively. They tried outcomes on every one of the 5014 phishing and 5000 non-phishing messages from Nazario and Enron Corpus email set. To extract all (verb-direct object) pairs from all sentences they used Stanford typed dependency parser and then AI is applied.

Subsequent to training, they considered an (verb-direct object) pair to be harmful if its sureness surpassed a threshold. They explored different avenues regarding diverse certainty shorts to tradeoff the requirement for high precision and for low false-positive rate. If the score is higher than 0.9 then the pair is malicious. They have assessed the test corpus with SEAHound algorithm and with Netcraft alone, which just identifies phishing URL links. The algorithm uses python language and implemented on an Intel Core i7 processor.

They compare SEAHound and Netcraft algorithm with five qualities for each methodology, true positives (TP), false positives (FP), false negatives (FN), precision, and recall. Precision is calculated by TP/TP+FP. Recall is calculated by TP/TP+FN (Figure 9.2).

Figure 9.2 demonstrates that this methodology, when contrasted with Netcraft, results in a decreased number of FP to the detriment of the quantity of FN. Decline in FN demonstrates that semantic information is a valuable indicator to recognize phishing assaults.

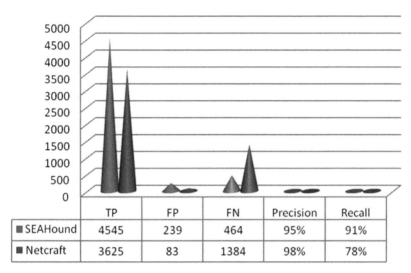

	TP	FP	FN	Precision	Recall
■ SEAHound	4545	239	464	95%	91%
■ Netcraft	3625	83	1384	98%	78%

FIGURE 9.2   Shows phishing detection results.

9.2   RELATED WORK REGARDING DETECTION OF PHISHING WEBSITES

Blacklists and Machine Learning are two methodologies described by the writers in this paper [15] which can be used to recognize phishing. In Blacklists, some blacklist suppliers were referenced and in machine learning Google Safe Browsing API, DNS Based Blacklist, and Phisnet. Some blacklist suppliers are referenced in Blacklists. And Google Safe Browsing API, DNS Based Blacklist, and Phisnet are mentioned in machine learning methodologies. Some of the other algorithms are additionally referenced where a short presentation is described about the finding the phishing location using machine learning. Two or three systems (Cybersquatting and typosquatting) were quickly examined which are regularly utilized to control URL by the attackers who are using phishing. At that point, a portion of the highlights of sites were viewed as which may characterize the site as phishing or not. In spite of the fact that, there was no notice of any of predictions after use of AI calculations and no reasonable procedure about component extraction was referenced.

In this paper, the writers describe a neural system as a technique to distinguish phishing sites [16]. Supervised learning algorithms (LAs) were utilized specifically Adaline network, Backpropagation system, and support vector machine (SVM). The authors applied these algorithms autonomously. And furthermore, they have applied blend of these algorithms. The primary objective are to preprocess the data, apply cleansing process to the data and then to extract the desired features of the site. They have extracted around 15 different features of the sites under consideration. Most noteworthy precision was seen by utilizing Adaline network with SVM. And they have considered more than 30 characteristics and their extraction was done through Python.

In another paper, they proposed to detect phishing websites through some of the many features that can be extracted from a URL [17]. After extracting those features and examining

Them, they decide whether a website is a phishing website or a legitimate one. They then came up with a simple module to find out if the web site is phishy or a legitimate one. This module uses the URL to carry out its task. The dataset they use consists of only about 100 URLs from the database of the Phistank and Yahoo directories. Of those 100 URLs, 59 have been legitimate and 41 have been phishing. The authors are able to achieve 98.4% of accuracy in deciding a website as phishing or legitimate with their proposed method.

For detecting phishing sites, the authors in a paper suggested the use of machine- LAs [18]. For this purpose, they have used public key certificates from X.509. They have referred PhishTank entries and used the certificates of those sites listed in the entries. Those certificates are examined even if HTTPS was included in the URL. Some of the characteristics like NotBefore, NotAfter, Issuer, Subject, Date-Downloaded, and Domain Name, etc. are used. Python is used to extract the features. Different machine learning techniques applied by them are-Decision Trees (DT), random forest (RF), Naïve Bayes, and logistic regression (LR). Form the methods they have used RF technique has performed the best and predicted with an accuracy of 95.5%.

PhishTank and OpenPhish were the sources of phishing data set in this paper [19]. The URL used for phishing, IP, brand name of the target, etc. are the features included in these datasets. Previously J48, LR, and SVM are the algorithms for this purpose. They have observed 96.96% of accuracy with J48 which is the best one. The authors used different datasets, some other machine LAs, and different techniques for pre-processing of data sets. The features of the URLs are also extracted using Python in order to identify it as Legitimate or Phishing. And with these modifications, they have achieved better precision.

Their proposal evaluates the association among the words which make up a URL [20]. The authors make use of the data used for querying in a search engine. They try to establish the relations between the words. They have suggested that these related words are better option for the vocabulary on the Internet as compared to other methods those have been used so far.

In the first step, they try to identify URL obfuscation. It can be performed by using keywords, domain name, IP address, and relative URL. In the next step, it prepares group of terms which are associated with domain names and then the other part of the URL is formed. These groups of words act as input sets for different algorithms used. 12 characteristics obtained from those sets. Machine learning techniques such as SVM, RF, and Random Tree are applied on those extracted characteristics. They have obtained the best accuracy with RF and that is 95.22%. They have the scope of improvement as they have only considered English vocabulary, whereas in the Internet lots of languages are used.

A predictive method is proposed by the authors in Ref. [21]. This method is based on characteristics of URL and ranks associated with the sites. The URL used to calculate Alexa Reputation. They have calculated root mean squared error to evaluate the correctness of different values. This technique's correctness is 97.16% with a 0.4 threshold. Their shortcomings were important features such as customization of the status bar, submission of information, neglected website forwarding, which could lead to incorrect prediction.

In [22] they have taken out website URL characteristics. Then they have used subset-based feature selection methods for analyzing those features. Based on these observations they proposed algorithms to detect phishing websites based on classification. They first determined the most targeted brand names and their legit URL from the PhishTank website via Google and the URLs which are used for phishing. They have defined 133 features matrix. Then a feature selection technique is used to reduce it. Naïve Bayes and sequential minimal optimization (SMO) are the classification algorithms applied to classify the dataset. Out of them, SMO classifier yielded the best result with 95.39%. They have the scope of improvement with respect to the percentage of accuracy.

9.2.1   PHISHING DETECTION METHODOLOGY

Phishing detection methodologies are illustrated in Figure 9.3.

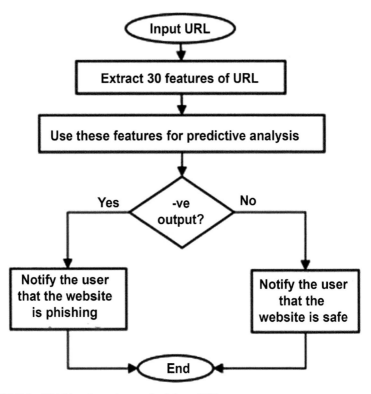

FIGURE 9.3   Phishing detection methodology [38].

9.2.2   BASIC PRINCIPLE STEPS REQUIRED TO RESOLVE THE WEBSITE PHISHING PROBLEMS

To resolve the phishing problem the following basic steps are stated as below [42].

- **Step 1:** Identification of desired information: It needed a cluster of attributes due to any kind of problem, which predetermined and desired classifier results. Therefore, a group of features for input as well as output and should be well informed.

- **Step 2:** Training dataset: It includes examples of inputs and preferred target attributes. There are several ways to obtain the phishing dataset, such as PhishTank; in addition, the phishing website dataset can be obtained on the UCI repository.
- **Step 3:** Choose the classification algorithm: It is a very difficult phase to choose the data mining (DM) algorithm. In the literature, there are many methods and techniques for DM where each method and technique has its own benefits and deficits.
- **Step 4:** Performance evaluation of different classifiers: The last step is to verify the overall performance and efficiency assessment of the determined classifier with regard to test data.

9.2.3   LIST OF MACHINE LEARNING ALGORITHMS (LAS) USED TO DETECT A WEBSITE GENUINE OR PHISHY

1. **Decision Tree (DT):** This algorithm is used to estimate the information gain for all features according to its entropy before adding a node (feature) to the DT. Based on the highest value of a feature; it will generate a single decision. Then the selection of the feature is iterated on the other features. It divides the highest value on the feature. The terminal value is the mode of observation. Thus, the mode value is used to predict new data. The user defines the depth of the tree.

2. **Random Forest (RF):** This algorithm adds randomness to DTs generation. Here it will not rely upon a single DT to cover all the features of the dataset. It randomly chooses training data and features from a given dataset. Basing upon the randomly selected inputs, it constructs series of DTs. RF's output is computed by DTs outputs encompassed there in. Each tree provides a classification for a new data. The forest has the most votes to choose the classification.

3. **Generalized Additive Model (GAM):** This system works by combining two different models to create a single model. GLM & RF combined gave the best result. At first, these two algorithms are applied individually to fit the model. Then GAM is applied to these algorithms 'combined results.

4. **Gradient Boosting (GBM):** To enhance model fitting boosting with machine LAs are used. To make strong predictors, it takes many weak predictors; add weight to the predictors. Based on errors, weight is calculated. It increases the weight of missed

classifications, thus transforming weak learners into strong learners. Tress boosting is called gradient boosting (GBM).

9.2.4 FACTORS INVOLVED IN CLEANING AND PRE-PROCESSING OF DATASET

1. **Principal Component Analysis (PCA):** Here highly correlated variables are combined using suitable combinations. It therefore produces novel variables in the removal of highly correlated variables in the dataset. This newly constructed dataset gave all the algorithms the best prediction.
2. **Variability Inflation Factor (VIF):** VIF is the increase in the variance of ith regressor compared to an ideal setting when it is orthogonal. This variability occurs because the variable is highly correlated with other variables apart from the result variable. The high VIF value variables were eliminated from the dataset, which further augmented the prediction rate of the result.
3. **K-Nearest-Neighbors (KNN):** KNN algorithm is based on an instance. For each unknown instance, a majority vote of the K training instances closest to that instance (based on the characteristics) determines its category. On the dataset, the KNN algorithm was used to extract values from attributes that were hard to find despite the limited computing resources. Those attributes were: These attributes were: number of links trying to point to a page, features Server Form Handler (SFH), and features based on statistics.

The author [38] applied different machine LAs on dataset. Generalized additive model (GAM) accuracy was 74% and generalized linear model (GLM) accuracy was 93.33%. Then, to increase the accuracy, some methods were used. They computed the accuracy for three top algorithms after eliminating the attributes with high VIF values. The accuracy of the RF algorithm after PCA was up to 98.4%.

9.3   INTRODUCTION TO HYBRID MODEL

The model helps to determine the phishing site and legit sites with the mixture of K-Means Clustering and Naive Bayes (NB) Classifier [24] based on the site's URL features [23] and HTML features. Website's URL features are applied with K-Means Clustering. Then the feature sets are plotted into three different database clusters. If the feature set is closed to valid phish then website is treated as phishing website. If the feature set is closed to invalid phish then website is treated as genuine website. A site is treated as suspicious site if feature set is not closer to valid and invalid phish. NB Classifier will be used if K-Means clustering technique is not effective.

9.3.1 SYSTEM ARCHITECTURE AND PROCEDURE

Figure 9.4 represents the system architecture [40] is given below:

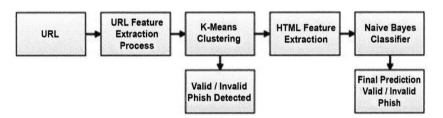

FIGURE 9.4   Represents system architecture.

The system architecture consists of four main parts.

Procedure:

The procedure for the above architecture is described as follows:

- **Stage 1:** Initially website W is given.
- **Stage 2:** In the second step uniform resource locator, feature extraction process occurs which extract the URL features from the website W.
- **Stage 3:** In this stage K-means clustering algorithm is applied on dataset of W. It will predict the cluster where W is closer to the centroid. The 0 value represents the W is suspicious. The value +1 and −1 represents the W is phishing and legit site respectively.
- **Stage 4:** If the value is +1 or −1 then it will predict the result otherwise it will move to stage number 5.
- **Stage 5:** In HTML feature extraction stage, HTML tag features are extracted and enter into W.

- **Stage 6:** In the last phase the website, W is classified with the help of Naïve Bayes classifier and it will predict the website is a valid or invalid phish.

9.3.2 URL FEATURE SET

The model's main function depends on what attributes to detect the phishing attack in the dataset. To identify the phishing attacks four features are chosen from W3C standards that can efficiently identify the phishing attacks:

1. **Dots in URL:** The dot in the URL implies the sub domain's presence. More the dots present in the URL, more the sub-domain in the URL to obstruct the URL of the web and look the genuine website alike. Some phishers may utilize the subdomain to look at the address of the site as the genuine site, making the client misdirect the site to phish.

2. **IP Address in URL:** The domain must be registered to acquire a particular URL address. Genuine websites register their domain area and have the address of the URL. This will enable us to recognize the phishing site. Phishing websites keep going for a couple of days, so the phisher may not enlist in the site.

3. **Slashes in URL:** The slash represents a subfolder, and cannot be replaced by any other symbol. To hide the data in the site pages, the subfolders are included.

4. **Suspicious Characters:** To trick the user, the phisher will use some other special characters than alpha-numeric characters. Unique characters in the web URL may have utilized ' @,"&,"- 'and' to make the pattern of the genuine URL that the user can easily click on.

9.3.3 K-MEANS CLUSTERING TECHNIQUE

The author [40] going through 1000 records which are present in the repository (www.phishtank.com) [8]. The phish and legit URL have +1 and –1 values (Figure 9.5).

Thus, with the clustering application, they can create the database. The database on the characteristics can be divided into three clusters as valid, invalid, and suspicious phish. The cluster includes feature sets with high and low values treated as valid and invalid phish respectively. If the features contain some sites as valid and some sites as invalid, then this cluster represents suspicious phish.

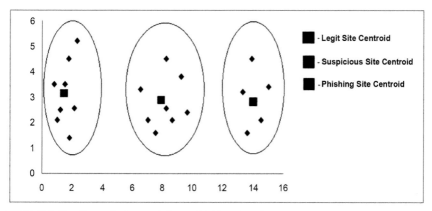

FIGURE 9.5   Shows legit, suspicious, phishing sites feature set.

For our clustering k = 3, K-Means Clustering refers to the clustering of the database of n feature sets using k partitions. For the initial clustering, initial values for clusters must be provided [25]. The centroid values are updated in every cycle to measure the distance between each centroid and feature set.

fs1, fs2, fs3... fsn represents set of features consisting of 4-dimensional vectors, K-Means clusters with the below equation:

$$\arg\min \sum_{i=1}^{\partial} \sum_{x_j \in s_i} \left\| x_i - u_j \right\|^2 \tag{1}$$

9.3.4   NAIVE BAYES (NB) CLASSIFIER

In the training data, Bayesian classifiers find the attribute values distribution for each class. Bayesian classifiers use Bayes theorem [25] which is used to find probability $p(C_j|d)$ of the instance d being in class C_j represented in equation 2.

$$p(C_j \mid d) = \frac{p(d \mid c_j)p(C_j)}{p(d)} \tag{2}$$

NB represented in Equation 3.

$$V_{nb} = \arg\ max_{fs_i \in V} P(f s_i)\ \Pi P(f s_i \mid C_j) \qquad (3)$$

Equation (4) represents the NB classifier which estimates $p(f_s|C)$ using m-estimates [9]. n_c denotes number of examples for which $f_s = fs$ and $C = C_j$. m denotes arbitrary value, equivalent sample size. n denotes number of training examples for which $fs_i = fs$. p denotes 1/number of values of an attribute.

$$p(f\ s_i \mid C_j) = \frac{n_c + mp}{n + m} \qquad (4)$$

NB classifier is used over the DT classifiers because they can classify estimations of invalid traits by overlooking them from probability computation. NB classifier cannot concretely handle null or unknown attributes, so that results will be more exact as compared to DT classifiers.

According to the author [26] out of 1000 records, 600 are valid and rest legit sites. The training set will be useful to predict the outcome using NB classifier.

Accordingly, the author [40] concludes that the Bayesian approach generates more accurate results. The Bayesian approach requires takes a long time to complete. K-means clustering technique that is efficient to generate output at higher performance but it lacks efficiency and is recovered by the help of NB Classifier.

9.4   USING SUPERVISED LEARNING ALGORITHMS (LAS) TO DETECT PHISHING-SITES

A hybrid model-based methodology has been used to take care of the phishing sites issues. A single model can't detect phishing websites efficiently because the single model needs to be enhanced. Enhancement in single model can be done by combining any two or three models to improve the accuracy of phishing site attack detection. The procedure of hybrid model is displayed in Figure 9.6 [39].

Sets of data collected from the source mentioned in [27] on the phishing website. A dataset is publicly available on the UCI repository. 30 features got selected from the phishing website.

DT, Bayesian net (BN), RF, SMO, NB, Instance-based learning (IBk) and fuzzy unordered rule induction (FURIA) classifiers are provided with dataset as training and testing. The job of above classifiers is to evaluate accuracy. The main approach of hybrid model is to checks the individual classifier's performance (in terms of less error rate and high accuracy) and gets the best classifier. Here the author combines other classifier models with the best classifier model for generating a hybrid model with high accuracy and less error rate.

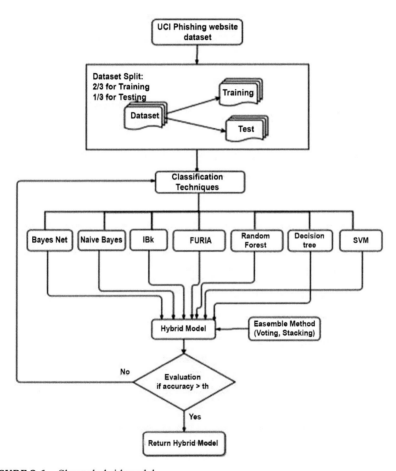

FIGURE 9.6 Shows hybrid model.

1. **Dataset:** The author used a publicly available data from the UCI repository [27]. Dataset is made up of 11055 instances with 30 attributes.

2. **Data Splitting Criteria:** Here data is divided into 2:3 ratios as training and testing.
3. **Data Mining Classification Techniques:** Here DT [29], RF [28], FURIA [31], SMO [30], IBk, NB, and BN algorithms classifiers are used to additionally examine and evaluate the relationships of the distinct characteristics of phishing classification. The tests are completed in netbeans utilizing the Weka API [32] Java language to conduct hybrid model classification experiment.
4. **Ensemble Methods:** This combines the base predictions of several estimators constructed by an algorithm and enhances robustness and generality over a single estimator [33].
5. **Hybrid Model:** Multiple models [34] incorporate in this phase to achieve higher accuracy and better performance. Hybrid model eliminates the individual model's drawbacks and takes the best features of two or more model in order to achieve higher accuracy and better performance. Here each model is combined with other models (model 1, model 2…) to form a hybrid model with some refinements. Pedro Domingo's [35] proposed the hybrid model. The pseudo-code of the hybrid model stated as below:

Inputs:

DS (training dataset)
LA (Learning algorithm)
MC (model combination procedure)
m(numbers of model to generate)
n (number of new examples to generate)

Procedure CMM (DS, LA, MC, m, n)

For i = 1 to m
DS_i (variation of DS)
Let m_i = produced model by applying LA to DS_i
For j = 1 to n
a (randomly generated example)
c (class assigned to a by $MC_{M1.Mm}$)
DS = DS U {(a, MC)}
Model = produced model by applying LA to DS
Return Model.

Performance Evaluation of Hybrid Model

Performance is measured by the following parameters:
Error Rate is measured by the following formula:
(FP+FN)/(TP+TN+FP+FN)
F-Measure is evaluated by the following formula:
2[(Precision*Recall)/[(Precision+ Recall)]
Accuracy is evaluated by the following formula:
(TP+TN)/(TP+TN+FP+FN)
Confusion Matrix: The Table 9.1 represents the confusion matrix.

TABLE 9.1 Confusion Matrix

Actual vs. Predicted	Positive(P)	Negative(N)
Positive(P)	TP	FN
Negative(N)	FP	TN

9.5 ANALYSIS OF HYBRID MODEL

The main goal of the hybrid model is to improve weak classifier and formed a hybrid classifier with the help of the best classifier so that it will provide higher accuracy and with less error rates. It is simulated using Weka API to classify phishing sites in netbeans java implemented code. It accurately detects genuine and phishy websites.

The author used several classification techniques (RF, Naïve Bayes, Bayes net, SMO, FURIA, IBk, DT) to identify genuine and phishing sites. The accuracy of the classification model is calculated on the data set of training and testing.

IBk and RF, were more than 95% accurate, showing the efficient models, although RF gained maximum classification accuracy 97.58% as per test data. Individual classification performances are depicted in Figure 9.7. RF classification technique is best model among all other techniques.

Additional performance methods like recall, f-measure, error rate, and accuracy are evaluated by the formula described in this chapter. The performance measure of RF classification techniques is depicted in Figure 9.8.

In phase 2, the author combines various classification models for efficient phishing site detection with improved accuracy even when combining models to improve the accuracy of individual models. According to their research SMO+IBk, RF+IBk, NB+IBk, BN+IBk, +IBk, J48+RF, NB+RF, BN+RF, and SMO+RF achieve the best result compared to achieved in individual models Figure 9.9 shows the classification accuracy of combined classifier techniques. A combination of DT+IBk and BN+IBk gives the highest precision in test data.

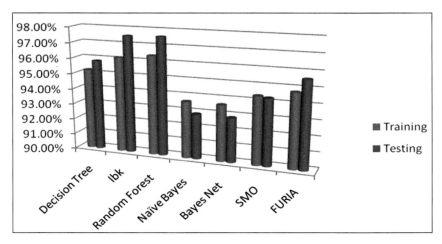

FIGURE 9.7   Performance of different classification techniques.

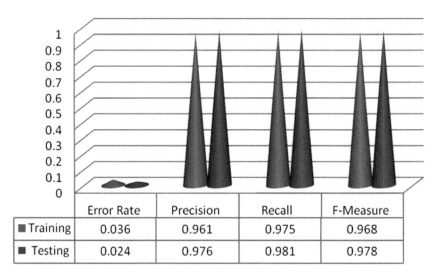

FIGURE 9.8   Performance measure of random forest classification technique.

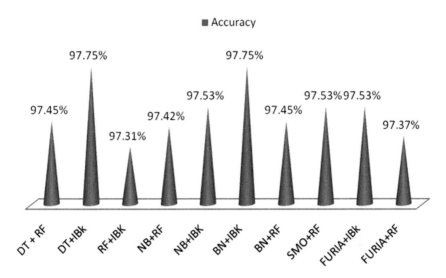

FIGURE 9.9   The accuracy of the combined classifier model.

The performance measure of the best hybrid model (DT+IBk) and (BN+IBk) is illustrated in Figure 9.10. The accuracy of the above two models is 97.75%.

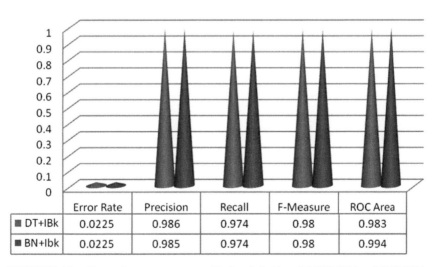

	Error Rate	Precision	Recall	F-Measure	ROC Area
■ DT+IBk	0.0225	0.986	0.974	0.98	0.983
■ BN+Ibk	0.0225	0.985	0.974	0.98	0.994

FIGURE 9.10   The performance measure of the best hybrid model (DT+IBk) and (BN+IBk).

9.6 CONCLUSION

The phishing website has been one of the global security challenges since the last few decades. It is a very challenging job to detect websites as genuine or phishy. This book chapter focuses not on metadata associated with emails, but on text analysis. As a result, phishing emails comprised of pure text are detected effectively. This chapter of the book consists mainly of machine learning techniques for detecting phishing websites. For results that are more accurate Bayesian, the approach is advisable. This approach requires a lot of time to complete the process because it analyzes the training data. K-Means Clustering that is effective to generate output at higher performance but is regained with the NB Classifier due to lack of efficiency.

Individual classification techniques are performed in Phase 1 of the hybrid model and the best three best models are selected based on performance criteria and high precision. In Phase II, each model is combined with the best three individual models of the hybrid model. The hybrid model provides accurate results as compared to the individual classification model. In testing data, the hybrid model achieved maximum accuracy of 97.75% as a combination of BN with IBk model and equal maximum accuracy on both hybrid models when DT ensemble with IBk model and less error rate of 0.225.

KEYWORDS

- Bayesian net
- hybrid model
- learning algorithm
- machine learning
- natural language processing
- phishing attacks

REFERENCES

1. Liu, P., & Moh, T. S., (2016). "Content-based spam e-mail filtering." In: *2016 International Conference on Collaboration Technologies and Systems (CTS)* (pp. 218–224). Orlando, FL.

2. Agrawal, N., & Singh, S., (2016). "Origin (dynamic blacklisting) based spammer detection and spam mail filtering approach." *2016 Third International Conference on Digital Information Processing, Data Mining, and Wireless Communications (DIPDMWC)* (pp. 99–104). Moscow.

3. Greitzer, F. L., Strozer, J. R., Cohen, S., Moore, A. P., Mundie, D., & Cowley, J. (2014). Analysis of Unintentional Insider Threats Deriving from Social Engineering Exploits. 2014 *IEEE Security and Privacy Workshops.*

4. Engin, K. C. K., (2005). "Protecting users against phishing attacks with anti phish." In: *Annual International Computer Software and Applications Conference.* Vienna.

5. Zhang, Y., Egelman, S., Cranor, L., & Hong, J., (2011). *"Phinding Phish: Evaluating Anti-Phishing Tools."* in IEEE.

6. Bitdefender Traffic Light, (2002). *BitDefender.* [Online]. Available: https://www.bitdefender.com/solutions/trafficlight.html (accessed on 26 February 2020).

7. Google Chrome, (2019). *"Chrome Web Store."* [Online]. Available: https://chrome.google.com/webstore/detail/phishdetector-true-phishi/kgecldbalfgmgelepbblodfoogmjdgmj (accessed on 26 February 2020).

8. A.P.W. Group, (2019). *"Phishing Activity Trends Report."* [Online]. Available: https://docs.apwg.org//reports/APWG_Global_Phishing_Report_2015-2016.pdf (accessed on 26 February 2020).

9. Stone, A., (2007). Natural-language processing for intrusion detection. *Computer, 40,* 12.

10. Verma, R., Shashidhar, N., & Hossain, N., (2012). Detecting phishing emails the natural language way. In: Foresti, S., Yung, M., & Martinelli, F., (eds.), *Computer Security ESORICS 2012 of Lecture Notes in Computer Science* (Vol. 7459). Springer Berlin Heidelberg.

11. Cranor, L., Egelman, S., Hong, J., & Zhang, Y., (2006). *Phinding Phish: An Evaluation of Anti-Phishing Toolbars.* Tech. Rep. CMU-CyLab-06-018, Carnegie Mellon University Cy Lab.

12. Pedregosa, F., Varoquaux, G., Gramfort, A., Michel, V., Thirion, B., Grisel, O., et al., (2011). Scikit-learn: Machine learning in Python. *Journal of Machine Learning Research, 12,* 2825–2830.

13. Park, Gilchan, & Julia M. Taylor (2015). "Using syntactic features for phishing detection." *arXiv preprint arXiv:1506.00037.* (accessed on 26 February 2020).

14. Klimt, B., & Yang, Y., (2004). *The Enron Corpus: A New Dataset for Email Classification Research.*

15. Ebubekir, B., Önder, D., & Ozgur, K. S., (2017). "Feature selections for the machine learning based detection of phishing websites." In: *2017 International Artificial Intelligence and Data Processing Symposium (IDAP).*

16. Singh, P., Maravi, Y. P. S., & Sharma, S., (2015). "Phishing websites detection through supervised learning networks." In: *IEEE International Conference on Computing and Communications Technologies (ICCCT)* (pp. 61–65).

17. Ahmed, A. A., & Abdullah, N. A., (2016). "Real time detection of phishing websites." In: 7ᵗʰ *IEEE Annual Information Technology, Electronics and Mobile Communication Conference.* IEEE IEMCON.

18. Dan, D. Z., Kapadia, A., Blythe, J., & Camp, L. J., (2015). "Beyond the lock icon: Real-time detection of phishing websites using public key certificates" In: *IEEE APWG Symposium on Electronic Crime Research* (pp. 1–12).

19. Marchal, S., Francois, J., State, R., & Engel, T., (2014). "Phish Score: Hacking phishers' minds." In: *Proceedings of the 10ᵗʰ International Conference on Network and Service Management 2014 (CNSM 2014)* (Vol. 11, No. 4, pp. 458–471).

20. Luong, A. T. N., Ba Lam, T., Huu, K. N., & Minh, H. N., (2014). "A novel approach for phishing detection using URL-based heuristic." In: *IEEE International Conference on Computing, Management and Telecommunications (ComManTel)* (pp. 298–303).

21. Mustafa, A., & Nazife, B., (2015). "Feature extraction and classification phishing websites based on URL." In: *IEEE International Conference on Communications and Network Security (CNS)* (pp. 769–770).

22. Mohammad, R. M., Thabtah, F., & McCluskey, L. (2012, December). An assessment of features related to phishing websites using an automated technique. In *2012 International Conference for Internet Technology and Secured Transactions* (pp. 492-497). IEEE.

23. Xiaoqing, G., Hongyuan, W., & Tongguang, N., (2013). "An efficient approach to detect phishing web." *Journal of Computational Information Systems, 9*(14), pp. 5553–5560.

24. Haijun, Z., Gang, L., Tommy, W. S. C., (2011). Senior member, IEEE, and Wenyin Liu, senior member. *IEEE "Textual and Visual Content-Based Anti-Phishing: A Bayesian Approach"* (Vol. 22, pp. 1532–1546). IEEE Transactions.

25. Silberschatz, A., Korth, H. F., & Sudarshan, S. (1997). *Database System Concepts* (Vol. 5). New York: McGraw-Hill.

26. Phishtank.com. *The Online Valid Phish Sites Repository.* http://data.phishtank.com/data/online-valid.csv (accessed on 26 February 2020).

27. http://archive.ics.uci.edu/ml/datasets/Phishing+Websites (accessed on 26 February 2020).

28. Biau, G., (2012). "Analysis of a random forests model." *The Journal of Machine Learning Research, 13*(1), 1063–1095.

29. Bhargava, N., et al., (2013). "Decision tree analysis on j48 algorithm for data mining." *Proceedings of International Journal of Advanced Research in Computer Science and Software Engineering* (Vol. 3, No. 6).

30. Chang, C. C., & Lin, C. J., (2011). "LIBSVM: A library for support vector machines." *ACM Transactions on Intelligent Systems and Technology (TIST), 2*(3), 27.

31. Huhn, J. C., & Hullermeier, E., (2010). "An analysis of the FURIA algorithm for fuzzy rule induction." In: *Advances in Machine Learning I* (pp. 321–344). Springer Berlin Heidelberg.

32. https://sourceforge.net/projects/weka/ (accessed on 26 February 2020).

33. Rokach, L., (2010). "Ensemble-based classifiers." *Artificial Intelligence Review, 33*(1/2), 1–39.

34. Lim, T. S., Loh, W. Y., & Shih, Y. S., (2000). "A comparison of prediction accuracy, complexity, and training time of thirty-three old and new classification algorithms." *Machine Learning, 40*(3), 203–228.

35. Domingos, P., (1997). "Knowledge acquisition from examples via multiple models." In: *Machine Learning-International Workshop Then Conference* (pp. 98–106). Morgan Kaufmann Publishers, Inc.

36. Baykara, M., & Gürel, Z. Z., (2018). Detection of phishing attacks. In: *2018 6ᵗʰ International Symposium on Digital Forensic and Security (ISDFS)* (pp. 1–5). IEEE.

37. Peng, T., Harris, I., & Sawa, Y., (2018). Detecting phishing attacks using natural language processing and machine learning. In: *2018 IEEE 12*th *International Conference on Semantic Computing (ICSC)* (pp. 300–301). IEEE.
38. Tyagi, I., Shad, J., Sharma, S., Gaur, S., & Kaur, G., (2018). A novel machine learning approach to detect phishing websites. In: *2018 5*th *International Conference on Signal Processing and Integrated Networks (SPIN)* (pp. 425–430). IEEE.
39. Tahir, M. A. U. H., Asghar, S., Zafar, A., & Gillani, S., (2016). A hybrid model to detect Phishing-Sites using supervised learning algorithms. In: *2016 International Conference on Computational Science and Computational Intelligence (CSCI)* (pp. 1126–1133). IEEE.
40. Patil, R., Dhamdhere, B. D., Dhonde, K. S., Chinchwade, R. G., & Mehetre, S. B., (2014). A hybrid model to detect phishing-sites using clustering and Bayesian approach. In: *International Conference for Convergence for Technology-2014* (pp. 1–5). IEEE.
41. Sharma, H., Meenakshi, E., & Bhatia, S. K., (2017). A comparative analysis and awareness survey of phishing detection tools. In: *2017 2*nd *IEEE International Conference on Recent Trends in Electronics, Information and Communication Technology (RTEICT)* (pp. 1437–1442). IEEE.
42. He, M., Horng, S. J., Fan, P., Khan, M. K., Run, R. S., Lai, J. L., Chen, R. J., & Sutanto, A., (2011). "An efficient phishing webpage detector." *Expert Systems with Applications, 38*(10), 12018–12027.

CHAPTER 10

Role of Computational Intelligence in Natural Language Processing

BISHWA RANJAN DAS[1] and BROJO KISHORE MISHRA[2]

[1]North Orissa University, Baripada, India,
E-mail: biswadas.bulu@gmail.com

[2]GIET University, Gunupur, Odisha,
E-mail: brojokishoremishra@gmail.com

ABSTRACT

Natural language processing (NLP) is a subfield of artificial intelligence (AI) and a research area in the field of computer science recently. It is processed by the computer system and understands the concept which is given as text input and generates some meaningful results. There are many subfields of NLP like machine translation, information retrieval (IR), information extraction (IE), and question answering. There are different types of natural languages are available in India even if in worldwide labels like Hindi, Odia, Bengali, Marathi, French, Spanish, German, etc. But different language has a different grammatical structure and rules. Most of the Indian languages are morphologically rich language. The process of natural language is mean to understand by the machine properly then processed it carefully and generates another natural language in the errorless manner by applying some intelligent computing algorithm. As per the mathematician or computer scientist, without algorithm the process of natural language is meaningless and will not to be produce output in proper format. Intelligent Computing is a method, ability to learn knowledge and specific task from the data set across a specified domain or experimental observation. The role of intelligent computing having various computational algorithms like machine learning, deep learning are applied in various parts of NLP like machine translation, sentiment analysis, IR, extraction, etc.

10.1 INTRODUCTION

Natural language processing (NLP) is a subfield of computer science, artificial intelligence (AI) concerned with the interactions between computers and human (natural) languages, in particular how to program computers to understand, process, and analyze large amounts of natural language data. Two major things are there, one is natural language understanding (NLU) and natural language generation (NLG). NLP is purely based on grammar-based (rule-based) and statistical-based. In the early days, many language-processing systems were designed by hand-coding a set of rules, e.g., by writing grammars or devising heuristic rules for stemming. However, this is rarely robust to natural language variation. Since the so-called "statistical revolution" [11, 12] in the late 1980s and mid-1990s, much NLP research has relied heavily on machine learning.

The machine-learning paradigm calls instead for using statistical inference to automatically learn such rules through the analysis of large corpora of typical real-world examples (a *corpus* (plural, "corpora") is a set of documents, possibly with human or computer annotations).

Many different classes of machine-learning algorithms (LAs) have been applied to natural-language-processing tasks. These algorithms take as input a large set of "features" that are generated from the input data. Some of the earliest-used algorithms, such as decision trees (DTs), produced systems of hard if-then rules similar to the systems of hand-written rules that were then common. Increasingly, however, research has focused on statistical models, which make soft, probabilistic decisions based on attaching real-valued weights to each input feature. Such models have the advantage that they can express the relative certainty of many different possible answers rather than only one, producing more reliable results when such a model is included as a component of a larger system.

Systems based on machine-LAs have many advantages over hand-produced rules:

- The learning procedures used during machine learning automatically focus on the most common cases, whereas when writing rules by hand it is often not at all obvious where the effort should be directed.

- Automatic learning procedures can make use of statistical inference algorithms to produce models that are robust to unfamiliar input (e.g., containing words or structures that have not been seen before) and to erroneous input (e.g., with misspelled words or words accidentally omitted). Generally, handling such input gracefully with hand-written rules—or, more generally, creating systems of hand-written rules that make soft decisions—is extremely difficult, error-prone, and time-consuming.
- Systems based on automatically learning the rules can be made more accurate simply by supplying more input data. However, systems based on hand-written rules can only be made more accurate by increasing the complexity of the rules, which is a much more difficult task. In particular, there is a limit to the complexity of systems based on hand-crafted rules, beyond which the systems become more and more unmanageable. However, creating more data to input to machine-learning systems simply requires a corresponding increase in the number of man-hours worked, generally without significant increases in the complexity of the annotation process.

10.2 STEPS OF NATURAL LANGUAGE PROCESSING (NLP)

NLP is done at five levels. These levels are briefly stated below:

- Phonology and phonetics;
- Morphological and lexical analysis;
- Syntactic analysis;
- Semantic analysis;
- Discourse integration;
- Pragmatic analysis.

10.2.1 MORPHOLOGICAL LEXICAL ANALYSIS

The "Lexicon" of a language is its vocabulary that includes its words and expression. "Morphology" is the identification, analysis, and description of the structure of words. The "words" are generally accepted as being the smallest units of syntax. The "syntax" refers to the rules and principles that given the sentence structure of any individual's language.

10.2.2   LEXICAL ANALYSIS

The aim is to divide the text into paragraphs, sentences, and words. The lexical analysis cannot be performed in isolation from the morphological and syntactic analysis.

10.2.3   SYNTACTIC ANALYSIS

Here the analysis is of words in a sentence to know the grammatical structure of the sentence. The words are transformed into structures that show how the words relate to each other. Some words sequences may be rejected if they violet the rules of the language for how words may be combined.

10.2.4   SEMANTIC ANALYSIS

It derives an absolute (dictionary definition) meaning from context; it determines the possible meaning of sentences in a context. The structures created by the syntactic analyzer are assigned meaning. Thus, a mapping is made between the syntactic structure and objects in the task domain. The structure for which no such mapping is possible is rejected.

10.2.5   DISCOURSE INTEGRATION

The meaning of an individual sentence may depend on the sentences that precede it and may influence the meaning of the sentence that follows it.

10.2.6   PRAGMATIC ANALYSIS

It derives knowledge from external common-sense information; it means understanding the purposeful use of language in situations, particularly those aspects of language which require world knowledge. The idea is what was said is reinterpreted to determine what was actually meant.

10.3   DEFINING TERMS RELATED TO LINGUISTIC ANALYSIS

The following terms are explained below:

- phone;
- phonetic;
- phonology;
- string;
- lexicon;
- words;
- determiner;
- morphology;
- morphemes;
- syntax;
- semantics;
- pragmatics;
- phrase;
- sentence.

1. **Phones:** The phones are acoustic patterns that are significant and distinguishable in some human language.
2. **Phonetics:** Tells how acoustic signals are classified into phones.
3. **Phonology:** Tells how phones are grouped together to form phonemes in particular human languages.
4. **Strings:** An alphabet is a finite "set of symbols" or a combination of symbols or sequences of symbols taken from alphabets.
5. **Lexicon:** It is collection of information about the words of a language. The information is about the lexical categories to which words belong. Lexical categories mean noun, verb, adjective, adverb, etc.
6. **Word:** It is a unit of language that carries meaning.
7. **Determiner:** It occurs before nouns and indicates the kind of reference which the noun has
8. **Morphology:** It is the analysis of words into morphemes and conversely the synthesis of words from morphemes.
9. **Morphemes:** A smallest meaning unit in the grammar of languages. A smallest linguistic unit that has semantic meaning. A unit of language immediately below the "word level." A smallest part of a word that can carry a discrete meaning. There are different types of

morphemes such as Free Morphemes, Bound Morphemes, Inflectional Morphemes, Derivational Morphemes, Root Morphemes, Null morphemes.

10. **Syntax:** It is the "structure of language." It is the grammatical arrangement of words in a sentence to one another in a sentence. Syntax rules govern proper sentence structure. Syntax is represented by parse tree, a way to show the structure of a language fragment or by a list.

11. **Semantics:** It is the meaning of words/phrases/sentences/whole text. Normally semantic is restricted to "meaning out of context," i.e., meaning as it can be determined without taking context into account.

12. **Pragmatic:** It tells "how language is used," that is "meaning in context."

13. **Grammatical Structure of Utterances:** Here sentence, constituents, phrase, classification, and structural rules are explained.

14. **Sentence:** It is a string of words satisfying grammatical rules of a language. Sentences are classified as simple, compound, and complex.

15. **Constituents:** Assume that a phrase is a construction of some kind. Here construction means a "syntactic arrangement" that consists of parts, usually two call constituents.

16. **Phrase:** It is a group of words (minimum is two) that functions as a single unit in the syntax of a sentence. It includes a noun phrase, verb phrase, and prepositional phrase.

17. **Grammar:** It is a declarative representation of syntactic facts about the language. It is the specification of the legal structures of language. It has three basic components i.e., terminal symbols, non-terminal symbols and production rules.

18. **Parser:** It is a procedure that compares the grammar against input sentences to produce a parsed structure called parse tree. A parser is a program that accepts as input a sequence of words in a natural language and breaks them into parts (nouns, verbs, and their attributes) to be managed by other programming. Parsing can be defined as the act of analyzing the grammatically an utterance according to some specific grammar. Parsing is the process to check that a particular sequence of words in a sentence corresponds to a language defined by its grammar. Parsing means show how we can get from

the start symbol of the grammar to the sequence of the words using the production rules. The output of the parser is a parse tree.

19. **Parse Tree:** It is way of representing the output of a parser. Each phrasal constituents found during parsing becomes a branch node of the parse tree. The words of the sentence become the leaves of the parse tree. There can be more than one parse tree for a single sentence.

20. **Context-Free Grammar:** In formal language theory, a context-free grammar is a grammar where every production rules are of the form A ——> α, where A is a single symbol called non-terminal and α is a string that is a sequence of symbols of terminals and/or non-terminals symbols. The terminal and non-terminal symbols are those symbols used to construct production rules in a formal grammar.

21. **Terminal Symbol:** Any symbol used in the grammar which does not appear on the left-hand side of some rule (i.e., has no definition) is called a terminal symbol. Terminal symbols cannot be broken down into smaller units without losing their literal meaning.

22. **Non-Terminal Symbol:** Symbols that are defined by rules are called a non-terminal symbol. Each production rule defines the non-terminal symbol, like the above rule states that "whenever we see an A, we can replace it with α."

23. **Regular Expression:** Every regular expression can be converted to a grammar, but not every grammar can be converted back to a regular expression. Any grammar which can be converted back to a regular expression is called a regular grammar the language it defines is a regular language.

24. **Regular Grammar:** It is a grammar where all of the production rules are of one of the following forms A ——> AB or A ——> a, where A and B represent any single non-terminal and 'a' represents any single terminal or the empty string.

10.4   COMPUTATIONAL INTELLIGENCE (CI)

The expression computational intelligence (CI) usually refers to the ability of a computer to learn a specific task from data or experimental observation. Even though it is commonly considered a synonym of soft computing, there is still no commonly accepted definition of CI.

Generally, CI is a set of nature-inspired computational methodologies and approaches to address complex real-world problems to which mathematical or traditional modeling can be useless for a few reasons: the processes might be too complex for mathematical reasoning, it might contain some uncertainties during the process, or the process might simply be stochastic in nature indeed, many real-life problems cannot be translated into binary language (unique values of 0 and 1) for computers to process it. CI therefore provides solutions for such problems.

The methods used are close to the human's way of reasoning, i.e., it uses inexact and incomplete knowledge, and it is able to produce control actions in an adaptive way. CI therefore uses a combination of five main complementary techniques. The fuzzy logic which enables the computer to understand natural language, artificial neural networks (NN) which permits the system to learn experiential data by operating like the biological one, evolutionary computing, which is based on the process of natural selection, learning theory, and probabilistic methods which helps dealing with uncertainty imprecision.

Except those main principles, currently popular approaches include biologically inspired algorithms such as swarm intelligence and artificial immune systems, which can be seen as a part of evolutionary computation, image processing, data mining (DM), NLP, and AI, which tends to be confused with CI. But although both CI and AI seek similar goals, there's a clear distinction between them.

There are few algorithms used in CI like support vector machine (SVM), artificial neural network, fuzzy logic, probabilistic method, and conditional random field, etc.

10.4.1 SUPPORT VECTOR MACHINE (SVM)

The SVMs are a binary learning machine with some highly elegant properties that are used for classification and regression. It is a well-known system for good generalization performance and it is used for pattern analysis. In NLP, it is applied to categorize the text, as it gives high accuracy with a large number of features set. We have used this machine to define a very simple case, a two-class problem where the classes are nonlinearly separable. Let the data set D to be given as (X_1,y_1), (X_2,y_2)….(X_D,y_D), where X_i is the set of training tuples with associated class labels y_i. Each y_i can take one of two values, either +1 or –1(i.e., $y_i \in \{+1,-1\}$ (Figure 10.1).

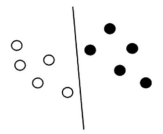 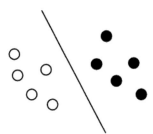

FIGURE 10.1 Classification of textual data.

A separating hyperplane equation can be written as wx + b = 0, where x is an input vector, w is the adjustable weight, and b is the bias. Training tuples are 2-D, e.g., x = $\{x_1, x_2\}$, where x_1, x_2 are the values of attributes A_1 and A_2 respectively for x. It finds an optimal hyperplane which separates the training data as well as the test data into two classes. It finds separating hyperplane which maximizes its margin. Two parallel lines and margin M can be expressed as wx + b = +1, M = 2/$\|w\|$. To maximizes this margins r = 1/$\|w\|$ and Minimize $\|w\|$ = $\|w\|^2/2$, Subject to d_i (w. x_i + b) ≥ 1, where i = 1, 2, 3,…, l). Any training tuples that falls on either side of the margins are called support vector. It has the strength to carry out the nonlinear classification. The optimization problem can be written usual form, where all feature vectors appear in their dot products. By simple substituting every dot product of x_i and x_j in dual form with a certain Kernel function K(x_i, x_j). SVM can handle nonlinear hypotheses. Among these, many kinds of Kernel function available. We shall focus on the polynomial kernel function with degree d such as K $(x_i, x_j) = (x_i * x_j + 1)^d$. Here d degree polynomial kernel function helps us to find the optimal separating hyperplane from all combination of features up to d. The hypothesis space under consideration is the set of functions. The linear separable case is almost done. The nonlinear SVM classifier gives a decision-making function f(x).

$$f(x)=\sum\nolimits_{i=1}^{m} w_i K(x,z_i)+b, \ g(x) = sign(f(x)) \tag{1}$$

If g(x) is +1, x is classified as class C_1 and –1 x is classified as class C_2. z_i are called support vectors and representative of training examples, m is the number of support vectors is a kernel that implicitly maps vectors into a higher dimensional space and can be evaluated efficiently. The polynomial kernel K(x, z_i) = $(x.z_i)^d$.

10.4.2   *ARTIFICIAL NEURAL NETWORK*

The functioning of neurons as information processing units in the human brain, so it is with NN made up of artificial neurons. The human nervous system may be viewed as a three-stage system (Figure 10.2).

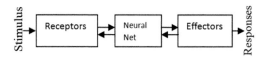

FIGURE 10.2   Function of neuron.

   Central to the system is the brain represented by the neural net, which continually receives information perceives it, and makes appropriate decisions. Two sets of arrows are shown left to right arrow indicate the forward transmission of information-bearing signals through the system. The arrows from right to left signify the presence of feedback in the system. Different classes of the neural network, i.e., Single layer feed-forward and Multilayer feed-forward network.

10.4.2.1   *SINGLE-LAYER FEED-FORWARD NETWORK*

Single-layer feed-forward network consists of a single layer of weights, where the inputs are directly connected to the outputs, via a series of weights. The synaptic links carrying weights connect every input to every output, but no other way. This way it is considered a network of feed-forward type. The sum of the products of the weights and the inputs is calculated in each neuron node, and if the value is above some threshold the neuron fires and takes the activated value otherwise it takes the deactivated value (Figure 10.3).

$$y_j = f(net_j) = 1 \text{ if } net_j \geq 0$$
$$0 \text{ if } net_j \leq 0 \text{ where } net_j = \sum_i x_i w_{ij}$$

10.4.2.2   *MULTI-LAYER FEED FORWARD NETWORK*

It consists of multiple layers. It has three layers, input, hidden layer & output layer. The computational units of the hidden layer are known as hidden neurons. A multi-layer feed-forward network with l input neurons, m1 neurons in the first hidden layers, m2 neurons in the second hidden layers, and n output neurons in the output layers is written as (l-m1-m2-n) (Figure 10.4).

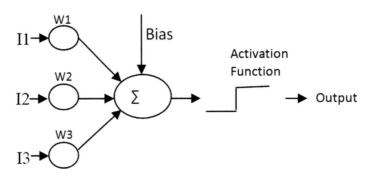

FIGURE 10.3   Function of single layer feed-forward network.

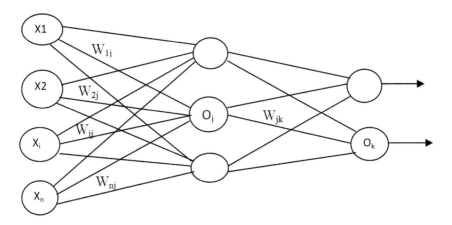

FIGURE 10.4   Function of multi-layer feed forward network.

$$x_1, x_2, x_3......x_n \text{ are the input.}$$

$$w_{k1,} w_{k2,} w_{k3}....w_{kn} \text{ are synaptic weight.}$$

$$U_k = \sum_{j=1}^{m} w_{kj} * x_j \tag{1}$$

$$Y_k = \varphi(u_k + b_k) \tag{2}$$

where;

    u_k = linear combiner output.

    $\varphi()$ = is the activation function.

    Y_k = Output signal of neuron apply an affine transformation to the output U_k of the linear combiner in the model

$$V_k = U_k + b_k \tag{3}$$

where; b_k is bias value which may +ve or –ve

The relationship between the induced local field or activation potential v_k of neuron k and the linear combiner output u_k is modified. The graph of v_k versus u_k no longer passes through the origin. The Bias b_k is an external parameter of artificial neuron k. I may account for its presence as in equation 2, equivalently; we may formulate the combination of equations (1) and (3) as follows.

$$v_k = \sum w_{kj}x_j \tag{4}$$

$$y_k = \varphi(v_k) \tag{5}$$

10.4.3   CONDITIONAL RANDOM FIELD

The conditional random field [4–6] is a model to calculate the conditional probability. CRF is an undirected graphical model which calculated the conditional probability of specified output node's values with given input node's values. CRFs defined the conditional probability distribution p *(S|O)* of the tagged sequence random variables set *S* on the observed sequence random variables set *O*, and finding the maximum probability *S** through the training to make *S** = *argmaxP(S|O)*`

The following equation is based on undirected graphical model:

$$exp\left(\sum_j \lambda_j t_j (y_{i-1}, y_i, x, i) + \sum_k \mu_k s_k (y_i, x, i) \right)$$

where; $t_j(y_{i-1}, y_i, x, i)$ is a transition feature function. $s_k(y_i, x, i)$ is a state feature function, λ_i and μ_k are parameters to be estimated from training data. At the time of defining features functions, a set of real-valued features b(x,i) of the observation is constructed to express some characteristic of the empirical distribution of the training data.

b(x, i) = {1 *if the observation at position i is the word anything*
0 *otherwise*

$t_j(y_{i-1}, y_i, x, i)$ = { b (x, i) if y_{i-1} = IN and y_i = NNP
0 otherwise

$$p(y|x, \lambda) = \frac{1}{z(x)} \exp(\sum_j \lambda_j f_j (y,x))$$

where; z(x) is the normalize factor.

10.4.4 FUZZY LOGIC

Fuzzy logic is a form of many-valued logic in which the truth values of variables may be any real number between 0 and 1 inclusive. It is employed to handle the concept of partial truth, where the truth value may range between completely true and completely false. By contrast, in Boolean logic, the truth values of variables may only be the integer values 0 or 1. The term fuzzy logic was introduced with the 1965 proposal of fuzzy set theory by LotfiZadeh. Fuzzy logic had however been studied since the 1920s, as infinite-valued logic—notably by Łukasiewicz and Tarski. It is based on the observation that people make decisions based on imprecise and non-numerical information, fuzzy models or sets are mathematical means of representing vagueness and imprecise information, hence the term fuzzy. These models have the capability of recognizing, representing, manipulating, interpreting, and utilizing data and information that are vague and lack certainty.

Classical logic only permits conclusions which are either true or false. However, there are also propositions with variable answers, such as one might find when asking a group of people to identify a color. In such instances, the truth appears as the result of reasoning from inexact or partial knowledge in which the sampled answers are mapped on a spectrum. Both degrees of truth and probabilities range between 0 and 1 and hence may seem similar at first, but fuzzy logic uses degrees of truth as a mathematical model of vagueness, while probability is a mathematical model of ignorance.

10.5   APPLYING TRUTH VALUES

A basic application might characterize various sub-ranges of a continuous variable. For instance, a temperature measurement for anti-lock brakes might have several separate membership functions defining particular temperature ranges needed to control the brakes properly. Each function maps the same temperature value to a truth value in the 0 to 1 range. These truth values can then be used to determine how the brakes should be controlled.

10.6   LINGUISTIC VARIABLES

While variables in mathematics usually take numerical values, in fuzzy logic applications, non-numeric values are often used to facilitate the expression of rules and facts.

A linguistic variable such as age may accept values such as young and its antonym old. Because natural languages do not always contain enough value terms to express a fuzzy value scale, it is common practice to modify linguistic values with adjectives or adverbs. For example, we can use the hedges rather and somewhat to construct the additional values rather old or somewhat young.

Fuzzification operations can map mathematical input values into fuzzy membership functions. And the opposite de-fuzzifying operations can be used to map a fuzzy output membership functions into a "crisp" output value that can be then used for decision or control purposes.

10.7   DIFFERENCE BETWEEN COMPUTATIONAL AND ARTIFICIAL INTELLIGENCE (AI)

Although AI and CI seek a similar long-term goal: reach general intelligence, which is the intelligence of a machine that could perform any intellectual task that a human being can; there's a clear difference between them. According to Bezdek [3] CI is a subset of AI.

There are two types of machine intelligence: the artificial one based on hard computing techniques and the computational one based on soft computing methods, which enable adaptation to many situations.

Hard computing techniques work following binary logic based on only two values (the Booleans true or false, 0 or 1) on which modern computers are based. One problem with this logic is that our natural language cannot always be translated easily into absolute terms of 0 and 1. Soft computing techniques, based on fuzzy logic can be useful here. Much closer to the way the human brain works by aggregating data to partial truths (Crisp/fuzzy systems); this logic is one of the main exclusive aspects of CI.

Within the same principles of fuzzy and binary logics, follow crispy and fuzzy systems. Crisp logic is a part of AI principles and consists of either including an element in a set, or not, whereas fuzzy systems (CI) enable elements to be partially in a set. Following this logic, each element can be given a degree of membership (from 0 to 1) and not exclusively one of these two values.

KEYWORDS

- **artificial intelligence**
- **computational intelligence**
- **fuzzy systems**
- **natural language generation**
- **natural language processing**
- **natural language understanding**

REFERENCES

1. https://en.wikipedia.org/wiki/Natural_language_processing (accessed on 26 February 2020).
2. https://en.wikipedia.org/wiki/Fuzzy_logic (accessed on 26 February 2020).
3. https://en.wikipedia.org/wiki/Computational_intelligence (accessed on 26 February 2020).
4. Ekbal, A., Haque, R., & Bandyopadhyay, S. (2007). "Bengali Part of Speech Tagging using Conditional Random Field." In Proceedings of the 7th International Symposium on Natural Language Processing (SNLP-07), pp. 131–136, Thailand.
5. Lafferty, J., McCallum, A., & Pereira, F. C. (2001). "Conditional random fields: Probabilistic models for segmenting and labeling sequence data," In Proc. of the 18th ICML'01, 282–289.
6. Hanna M. Wallach. (2004). "Conditional Random Fields: An Introduction," Technical Reports (CIS). Available at: http://works.bepress.com/hanna_wallach/1.

Index